AF899637

Contemporary Systems Thinking

Series Editors

Gianfranco Minati, Italian Systems Society, Milano, Italy

Alessandro Giuliani, Department of Mathematics and Physics, Università Roma Tre, Rome, Italy

Leonardo Bich, IAS Research Centre, University of the Basque Country, Leioa, Spain

Contemporary Systems Thinking is a series of texts focusing on interdisciplinary approaches, concepts, and issues of systems science and systems thinking (systemics). Texts in this series are not discursive but rather academic and scientific. Popular topics in the series include complexity, connectionism, chaos and emergence, as they are related to disciplinary systemics, theoretical systemics, applied systemics, and cultural systemics. Special consideration is given to texts about contemporary social issues that have predominant systemic aspects, such as climate, ecology, energy, public health, business, and others.

Fuad Aleskerov · Olga Khutorskaya ·
Anna Stepochkina · Vyacheslav Yakuba ·
Ksenia Zinovyeva

Bibliometric Analysis by Network Models

Identifying Trends in Scientific Literature

Springer

Fuad Aleskerov
International Centre of Decision Choice and Analysis
HSE University
Moscow, Russia

Laboratory of Choice Theory and Decision Analysis
Institute of Control Sciences of Russian Academy of Sciences
Moscow, Russia

Anna Stepochkina
International Centre of Decision Choice and Analysis
HSE University
Moscow, Russia

Ksenia Zinovyeva
International Centre of Decision Choice and Analysis
HSE University
Moscow, Russia

Olga Khutorskaya
Laboratory of Choice Theory and Decision Analysis
Institute of Control Sciences of Russian Academy of Sciences
Moscow, Russia

Vyacheslav Yakuba
International Centre of Decision Choice and Analysis
HSE University
Moscow, Russia

Laboratory of Choice Theory and Decision Analysis
Institute of Control Sciences of Russian Academy of Sciences
Moscow, Russia

ISSN 1568-2846
Contemporary Systems Thinking
ISBN 978-3-032-09170-3 ISBN 978-3-032-09171-0 (eBook)
https://doi.org/10.1007/978-3-032-09171-0

© The Editor(s) (if applicable) and The Author(s), under exclusive license to Springer Nature Switzerland AG 2026

This work is subject to copyright. All rights are solely and exclusively licensed by the Publisher, whether the whole or part of the material is concerned, specifically the rights of translation, reprinting, reuse of illustrations, recitation, broadcasting, reproduction on microfilms or in any other physical way, and transmission or information storage and retrieval, electronic adaptation, computer software, or by similar or dissimilar methodology now known or hereafter developed.
The use of general descriptive names, registered names, trademarks, service marks, etc. in this publication does not imply, even in the absence of a specific statement, that such names are exempt from the relevant protective laws and regulations and therefore free for general use.
The publisher, the authors and the editors are safe to assume that the advice and information in this book are believed to be true and accurate at the date of publication. Neither the publisher nor the authors or the editors give a warranty, expressed or implied, with respect to the material contained herein or for any errors or omissions that may have been made. The publisher remains neutral with regard to jurisdictional claims in published maps and institutional affiliations.

This Springer imprint is published by the registered company Springer Nature Switzerland AG
The registered company address is: Gewerbestrasse 11, 6330 Cham, Switzerland

If disposing of this product, please recycle the paper.

Competing Interests The authors have no competing interests to declare that are relevant to the content of this manuscript.

About This Book

The book contains new models in bibliometric analysis based on the new centrality measures in networks, pattern analysis and stability analysis. A distinctive feature of these centrality measures is that they take into account the parameters of vertices and group influence of vertices to a vertex. It allows to reveal a specific groups of publications, authors, terms, journals and affiliations of the authors depending on different parameters of publications. Pattern analysis and stability analysis allow to point out the tendencies in developing of the field of research over years. These new models are illustrated by an analysis of publications of 39,811 articles on various aspects of Parkinson's disease. We examined the attributes (3,292 journals, 27,551 authors, 3,029 affiliations, 423 terms) of publications published in 2015–2021. This methodology can be useful for researchers of any scientific domain, because it enables them to identify key and actively developing trends as well as major players in the field. Moreover, this approach allows to determine main tendencies in the entire research community as well as in the specific parts of it that may have gone unnoticed before. The obtained results of the analysis are useful not only for researchers but also for journals, editorial teams, scientific organizations, and investors.

Contents

Abbreviations

123i	radioactive isotope of iodine
a30p	alanine to proline at position 30 (mutation of SNCA gene)
AD	Alzheimer's Disease
API	Application Programming Interface
ASP	Article's Scientific Prestige
atp13a2	gene
BCG	Biochemistry, Cell, Genetics
BI	Bundle Index
CAG	Coronary AngioGram
cas9	protein 9
CNS	Central Nervous System
CRITIC	Criteria Importance Through Inter-criteria Correlation
CSV	Comma-Separated Values
CWTS	Centre for Science and Technology Studies (Netherlands: Centrum voor Wetenschaps- en Technologie Studies)
DBS	Deep Brain Stimulation
DEA	Data Envelopment Analysis
DOI	Digital Object Identifier
e3	protein
EEG	ElectroEncephaloGraphy
erk2	Extracellular Signal Regulated Kinase 2
fMRI	Functional magnetic resonance imaging
G2019S	mutation of LRRK2 gene
gba	gene Glucocerebrosidase
Guam	Guam disease
ISSN	International Standard Serial Number
JIF	Journal Impact Factor
JSON	JavaScript Object Notation
LDA	Latent Dirichlet Allocation
Lewy	Dementia with Lewy bodies
lrrk2	gene leucine-rich repeat kinase 2

MDS	Movement Disorder Society
MeSH	Medical Subject Headings
MJFF	Michael J. Fox Foundation
MMSE	The Mini-Mental State Examination (or Folstein test)
MPP	Mitochondrial Processing Peptidase
MRI	Magnetic Resonance Imaging
MSA	MultiSystem Atrophy
PD	Parkinson's Disease
PDD	Pervasive Developmental Disorders
PDQ	The Parkinson's Disease Questionnaire
PI	Pivotal Index
pink1	gene
RBD	REM Sleep Behavior Disorder
RCR	Relative Citation Rate
REM	Rapid Eye Movement
SCIE	Science Citation Index Expanded
SciMAT	Science Mapping Analysis Tool
SE	Social Entrepreneurship
SNA	Social Network Analysis
SNCA	SyNuClein Alpha gene
sy5y	cell models in PD research (SH-SY5Y)
System-GMM	System Generalized Method of Moments
TCS	Transcranial sonography
tDCS	Transcranial direct current stimulation
TI	Total Influence index
UPDRS	The Unified Parkinson's Disease Rating Scale
WHO	World Health Organization

Chapter 1
Introduction

For any researcher, there is a permanent need to obtain information on ongoing research worldwide. It is important to understand what has already been done, what is most relevant in the field of research, and to forecast the development in specific scientific areas.

Such information can be obtained using various knowledge databases, the volume of which increases annually. The traditional way for monitoring the development of scientific knowledge is the analysis of publications presented in scientific journals. When monitoring, it is important to record long-term changes in the thematic spectrum of scientific problems, i.e., to identify precisely the trends in various scientific directions.

One of the main tools that allows tracking global research activity is bibliometric analysis. Bibliometric analysis is a highly demanded and widely used method for measuring publication activity, monitoring the development of scientific directions, and the research activity of the scientific community. It allows to

- identify the most popular journals where studies on a particular topic are published,
- assess their impact and status within the scientific community,
- identify the most relevant research directions in a specific field,
- determine the countries and organizations leading in research on a particular topic,
- study cooperation between groups in the form of scientific publications, for example, a collaboration network between scientists and research institutions.

Thus, bibliometric analysis is a method for the analysis of trends and performance indicators of research in a particular field of knowledge.

The number of scientific publications (abstracts, articles, monographs, patents, etc.) is increasing every year and is currently estimated in petabytes, which requires specialized methods for processing accumulated information. Various forms of citation analysis are an important component of bibliometric analysis. According to

© The Editor(s) (if applicable) and The Author(s), under exclusive license to Springer Nature Switzerland AG 2026

F. Aleskerov et al., *Bibliometric Analysis by Network Models*, Contemporary Systems Thinking, https://doi.org/10.1007/978-3-032-09171-0_1

the definition by A. Prichard (Prichard, 1969), bibliometrics is "the application of mathematics and statistical methods to books and other media of communication."

The rapidly growing interest in network analysis has naturally led to the development of tools for research in this area. The exponential growth in the overall volume of publications underscores the relevance of tasks related to analyzing interconnections between scientific publications. In network science, models and methods are being developed to address these issues, falling within the domain of so-called citation networks.

This book deals with the analysis of bibliometric networks arising from the citation process of scientific articles indexed in scientific databases. Citation analysis methods can generally be divided into two classes: (1) citation statistics and (2) citation network analysis. The multifaceted nature of network analysis raises the task of combining network metrics to identify the most important publications. Network analysis is a powerful tool for analyzing complex interactions within research communities. We use a comprehensive approach to analyze publication activity based on network analysis methods. The objective of our work is to develop a methodology for studying a subject area using network models based on bibliometric data. The developed approach makes it possible to identify connections between research clusters, rank their significance, and track changes in the research directions of scientific groups.

Various network analysis methods are applied in this work to study citation patterns, and citation networks have been constructed for articles, journals, authors and their affiliations, and research topics. To determine the influence of particular entities, centrality indices have been used—four classical (Newman, 2018) and two new ones (Aleskerov & Yakuba, 2020). These indices are used to determine the importance of authors or publications within the scientific network, showing how they affect the overall scientific landscape. Unlike classical indices, the new centrality measures take into account vertex parameters, distant interactions between vertices, and group influence of vertices, thereby enhancing the understanding of how the analyzed networks function. The indices have been evaluated for the entire observation period as well as annually. For journals, citation pattern changes over the years are analyzed. The methodological tools for network research are important primarily for identifying promising areas of scientific inquiry. This work used not only citation networks and classical analysis tools but also other models and various analytical methods for a comprehensive study of the subject area. The new tools can be used in combination, allowing very new results.

To test the methodology we use scientific publications related to Parkinson's disease (PD). The following tasks have been formulated.

- analyze the dynamics of publications by year (for a given time period, namely 2015–2021) to determine how and in which directions research on Parkinson's disease is evolving;
- investigate the most productive and influential authors in this field;
- examine the number of citations for each publication and identify the most influential and cited works in the field;

- identify the most popular journals where research on Parkinson's disease is published;
- analyze the most frequently occurring keywords and topics in publications on Parkinson's disease.

Parkinson's disease and Alzheimer's disease are two scourges of the modern world and the two most common neurodegenerative diseases. More than ten million people worldwide suffer from Parkinson's disease (PD), and more than 30 million suffer from Alzheimer's disease (AD) (Olesen et al., 2012). According to the Parkinson's Foundation, about one million people in the United States are currently living with Parkinson's disease, and this number is expected to rise to 1.2 million by 2030 (Abeynayake & Tanner, 2020). Approximately 60,000 Americans are diagnosed with Parkinson's disease each year. In Canada, more than 100,000 Canadians live with Parkinson's disease, with approximately 6600 new cases diagnosed annually. In the United Kingdom, there were about 127,000 people with this disease in 2017. In 2022, the World Health Organization (WHO) reported that disability and mortality from Parkinson's disease are growing faster than from any other neurological disorder, including Alzheimer's disease. The disease is found throughout the world. In nearly all regions, the incidence of Parkinson's disease is increasing. Over the past 25 years, the number of people with PD has doubled, and experts predict it will double again by 2040 due to population aging.

The disease is a global issue not only of a social nature but also of an economic one. The economic consequences of the disease are catastrophic. The cost of the disease increases creating an economic burden for healthcare systems, society, and the patients themselves. Total cost estimates vary by country. The cost associated with Parkinson's disease in Europe amounts to €13.9 billion per year (Lindgren et al., 2005). The highest direct costs are typically associated with inpatient treatment and nursing home care. Costs rise from €5000 annually in the early stages of the disease to €17,000 or more in the late stages. Total costs in the United Kingdom are estimated at between £449 million and £3.3 billion annually, depending on the cost model used and the level of prevalence (Findley, 2007).

A study by the Lewin Group (The economic burden of Parkinson's disease, 2019), supported by the Michael J. Fox Foundation (MJFF) and the Parkinson's Foundation, showed that the total cost of PD to individuals, families, and the U.S. government is $51.9 billion annually. Of this, $25.4 billion is attributed to direct medical costs and $26.5 billion to non-medical costs such as missed work, lost wages, early forced retirement, and family caregiver time. In 2017, Parkinson's disease cost the United States approximately $52 billion per year, and it is projected that by 2037 this figure will increase to nearly $80 billion.

The significant socio-economic impact of Parkinson's disease due to the rising global incidence has led to a sharp increase in scientific interest in the issue. The number of studies has grown, and consequently, the number of publications on the topic has also increased. A rapid and representative analysis of publications has become highly relevant for the scientific and medical community to obtain information about the most significant and influential works, research teams, and current trends in studies. Such analysis allows the scientific and medical community to track the emergence of new research directions, assess their potential, and determine the influence of an author or research team on the perceptions of colleagues.

A developed comprehensive analysis of publication databases enables researchers to obtain information about key research areas, their development, discover research clusters and interconnections, identify key researchers, author affiliations, and journals. This approach also allows tracking the popularity dynamics of major research areas and drawing attention to previously unknown studies. The applied approach to the analysis of the scientific community is universal and can be used to analyze publication activity in various fields.

Our monograph deals with the development of these studies. About 40,000 publications related to a wide range of PD research in the period 2015–2021 have been reviewed. The monograph is divided into six chapters.

Chapter 2 provides a short survey of modern methods of bibliometric analysis of publications including those on PD. Next chapters also contain surveys of such publications related to the topic in the corresponding chapter.

Chapter 3 provides a description of the data obtained from the Microsoft Academic scientific publication database, the algorithm for preliminary analysis and filtering of articles. This chapter describes the citation networks discussed in the following chapters (publications, journals, authors, affiliations, and terminology). The chapter presents a theoretical description of network analysis methods, includes both classical and new centrality indices, examines network robustness analysis, pattern analysis, and community detection algorithms.

Chapter 4 focuses on the citation analysis of publications. It examines the citation network of publications and provides examples with calculated values of classical and new centrality indices. The dynamics of changes in the indices over the considered time period are analyzed.

Chapter 5 deals with the terminology analysis. The chapter includes the calculation of term frequency in the annotations of selected publications, grouping of terms by research directions, and analysis of frequency within a given direction. It describes the grouping of terms into pairs and triplets, and the analysis of these combinations both in the general term corpus and separately by research directions. Examples of results from term network analysis are provided for each research direction, along with the centrality analysis of publications associated with research directions based on the terms used.

Chapter 6 presents an assessment of the publication activity of journals using statistical and network analysis methods. It provides results of pattern analysis with various parameters.

Chapter 7 describes the citation analysis of publication authors and their affiliations. It includes results of statistical processing of author and affiliation citations, as well as the analysis of citation network indices for authors and affiliations. For affiliations, the results of citation network robustness analysis and pattern analysis are presented. This chapter also includes an analysis of the independent citation network of affiliations and the network of institutional collaboration. It presents the results of correlation analysis between the most highly cited affiliations and the most cited publication authors.

The new approaches to bibliometric analysis presented in the monograph provide more accurate and objective mechanisms for evaluating scientific activity and enable enhanced measurement of scientific impact.

Supplementary Materials Detailed evaluations are available on the website https://www.hse.ru/DeCAn/parkinson.

Acknowledgements F.Aleskerov, A.Stepochkina, V.Yakuba, K.Zinovyeva thank the Basic Research Program of the HSE University for partial financial support.

References

Abeynayake, I., & Tanner, C. M. (2020). The economic impact of OFF periods in Parkinson disease. *J Manag Care, 26*, 265–269. https://doi.org/10.37765/ajmc.2020.88518

Aleskerov, F., & Yakuba, V. (2020). Matrix-vector approach to construct generalized centrality indices in networks. *WP7/2020/01,* Moscow: Higher School of Economics Publ. House, SSRN. ssrn.com/abstract=3597948.

Findley, L. J. (2007). The economic impact of Parkinson's disease. *J Parkinsonism Relat Disord, 13*, 8–12. https://doi.org/10.1016/j.parkreldis.2007.06.003

Lindgren, P., von Campenhausen, S., Spottke, E., Siebert, U., & Dodel, R. (2005). Cost of Parkinson's disease in Europe. *European Journal of Neurology, 12*(1), 68–73. https://doi.org/10.1111/j.1468-1331.2005.01197.x

Newman, M. E. (2018). *Networks* (Second ed.). Oxford University Press.

Olesen, J., Gustavsson, A., Svensson, M., Wittchen, H.-U., & Jönsson, B. (2012). The economic cost of brain disorders in Europe. *European Journal of Neurology, 19*(1), 155–162. https://doi.org/10.1111/j.1468-1331.2011.03590.x

Prichard, A. (1969). Statistical bibliography or bibliometrics? *Journal of Documentation, 25*(4), 348–349.

The economic burden of Parkinson's disease. (2019). The Lewin group report. *date posted*, (July 5, 2019) www.parkinson.org/sites/default/files/documents/2019-parkinsons-economic-burden-study.pdf

Chapter 2
A Survey of the Network Bibliometric Analysis

Bibliometric analysis plays a crucial role in the systematic study of scientific publications. It is widely used to assess scientific productivity, identify key trends, sources of information, and thematic clusters. The main techniques of bibliometric analysis include citation analysis, co-citation analysis, and mapping of scientific networks, which allow for a detailed understanding of the interconnections between studies and their influence. The analysis of bibliometric indicators helps researchers make informed decisions regarding publication activity, funding allocation, and the development of scientific fields. While traditional bibliometric methods focus on quantitative indicators such as the number of publications and citation indices, network analysis provides a deeper understanding of how knowledge spreads and which institutions play leading roles in shaping scientific directions. Scientific network analysis enables the assessment of the impact of individual publications and the identification of scientific clusters, main research directions, and gaps in studied topics. Citation network analysis is a key tool for studying the structure of scientific research, identifying seminal works, and determining trends across various fields of knowledge. The use of centrality metrics helps to understand which nodes in the network have the greatest influence, how knowledge is disseminated within the academic community, and what factors may limit access to scientific information.

In this chapter, we provide an overview of some fundamental and recent works in the field of bibliometric analysis, not limited to those based on network models. Subsequent chapters will contain more focused literature reviews of specific research problems.

© The Editor(s) (if applicable) and The Author(s), under exclusive license to Springer Nature Switzerland AG 2026

F. Aleskerov et al., *Bibliometric Analysis by Network Models*, Contemporary Systems Thinking, https://doi.org/10.1007/978-3-032-09171-0_2

2.1 A Survey of Recent Works in Bibliometric Analysis

In (HabibAgahi et al., 2022), a co-authorship network in the field of Process Mining was analyzed, aiming to identify the structure of collaboration and key researchers. Centrality metrics like Closeness, Degree, Eigenvector, Betweenness, as defined in (Newman, 2018) were used to identify them. The TOPSIS method is used in the article. This multi-criteria decision-making method allow to rank researchers based on multiple influence indicators. The analysis, both with and without central nodes, made it possible to assess their influence on the network and showed that the network exhibits scale-free and small-world properties, with leading authors playing a key role in its connectivity.

In (Chen et al., 2023), the innovation network of the integrated circuit industry in China is analyzed. Using patent cooperation data from 2011–2020, the authors investigate how network analysis can explain cooperation mechanisms between cities and their influence on technological breakthroughs. Research methods include social network analysis to identify cooperation structures and the System-GMM method to reveal the evolutionary characteristics of networks. Results show that cities with high network centrality exhibit greater innovation activity, while universities and industrial zones amplify this effect by acting as institutional moderators. The study demonstrates that the innovative development of the integrated circuit industry depends not only on the network structure but also on the institutional environment that supports cooperation and technological advancement.

In (Minati, 2024) the concept of passive linkage in networks is introduced. According to the model additional arcs are superimposed on the actual existing network comprising the Multiple Superimposed Soft Network (MSSN). These additions could invoke extra information of the real links of the current network, including statistics, practices, temporal synchronizations, and other soft factors not represented by the weights of the original network. The MSSN model can be applied in different areas of research and especially benefits the machine-learning models.

Very important component of bibliometric studies lies in the use of different software. There exists a variety of software for the visualization and analysis of networks, such as *VOSviewer* (https://www.vosviewer.com/). *VOSviewer* is a specialized software tool developed at the Centre for Science and Technology Studies (CWTS) of Leiden University for constructing and visualizing bibliometric networks. The software supports data import from major bibliographic databases such as Web of Science, Scopus, PubMed, Dimensions, and OpenAlex. In addition to visualization, *VOSviewer* offers clustering functionalities to identify groups of related items within networks.

The *CiteSpace* software (https://citespace.podia.com/) is designed for the visualization and analysis of trends and patterns in scientific literature. The software facilitates the construction of co-citation networks, keyword co-occurrence networks, and collaboration networks among authors, institutions, and countries. *CiteSpace* calculates several centrality indices to assess the importance and influence of nodes in the constructed networks. While Betweenness is the primary centrality measure emphasized in *CiteSpace*, the software also enables the study of other network metrics such as Degree centrality and clustering coefficients using external tools.

The *R language* packages (*Biblioshiny*, *Bibliometrix*) (https://www.bibliometrix.org) include *Biblioshiny*, a web application based on the *Bibliometrix* R package. The software allows the import and processing of bibliographic data from various sources, including Web of Science, Scopus, PubMed, and Dimensions. The program supports co-occurrence analysis, co-citation analysis, and co-authorship network analysis. *Biblioshiny* computes various centrality indices in bibliometric networks, which are significant for understanding the influence and connectivity of nodes (e.g., authors, institutions, or countries) in scientific collaboration and citation networks. Centrality metrics available in Biblioshiny include Degree centrality, Closeness, Betweenness, Eigenvector, PageRank, Hub, and Authority Scores.

Bibliometrix supports a wide range of analytical methods, including descriptive bibliometric analysis, co-citation analysis, bibliographic coupling, thematic analysis, and collaboration network analysis (Aria & Cuccurullo, 2017). *Bibliometrix* integrates with *R*'s statistical and visualization tools, allowing for advanced analyses such as correspondence analysis, multidimensional scaling, and clustering.

SciMAT (Science Mapping Analysis Tool), developed at the University of Granada (https://sci2s.ugr.es/scimat/), is intended to facilitate comprehensive scientific analysis of publications. The software is capable of generating various types of bibliometric networks, including term co-occurrence, co-citation, bibliographic coupling, and co-authorship networks. The clustering algorithms available in *SciMAT* allow for the identification of thematic clusters within bibliometric networks. These clusters can be analyzed across sequential time periods to reveal the evolution of research topics. A distinctive feature of *SciMAT* is the use of strategic diagrams, which plot themes based on two dimensions: centrality and density. Centrality reflects the degree of interaction in the network, emphasizing importance within the broader research field. Density measures the internal cohesion of a network. In addition to centrality and density, *SciMAT* incorporates indicators such as the h-index, g-index, hg-index, and q^2-index to assess the impact and quality of research topics and clusters. *SciMAT*'s visualization capabilities include diagrams, thematic networks, and evolution maps.

CitNetExplorer is a specialized software tool designed for the visualization and analysis of citation networks in scientific literature. The program was developed by Nees Jan van Eck and Ludo Waltman at the Centre for Science and Technology Studies (CWTS) of Leiden University (https://www.citnetexplorer.nl/). A key feature of *CitNetExplorer* is its ability to import citation data directly from the Web of Science database, facilitating the creation of large-scale citation networks encompassing millions of publications and citation links. *CitNetExplorer* uses a clustering algorithm based on direct citation links to identify groups of closely related publications. The visualization component arranges publications along the horizontal time axis, while the vertical layout reflects citation relationships, thereby providing a temporal perspective on the development of scientific fields. *CitNetExplorer* focuses on the structural aspects of citation networks through clustering and visualization based on direct citation links. The program does not explicitly calculate traditional centrality measures such as Degree, Betweenness, or Closeness.

Pajek is a comprehensive software package designed for the analysis and visualization of large-scale networks containing up to one billion vertices (http://mrvar.fdv.uni-lj.si/pajek/). *Pajek* is used to construct and study networks representing relationships among scientific entities. For example, researchers can create two-mode networks linking authors to their publications or keywords to documents, enabling the identification of collaboration patterns and thematic structures. *Pajek* offers a suite of centrality indicators that quantitatively assess the importance or influence of nodes in a network. Key centrality indices computed by the software include: Degree centrality, Closeness, Betweenness, and Eigenvector. *Pajek*'s visualization tools complement its analytical capabilities, allowing users to generate graphical representations of networks.

Numerous other programs have been developed specifically for bibliometric analysis of scientific fields and for constructing maps of science. The development of specialized bibliometric analysis tools has significantly simplified the implementation of such studies. These tools allow not only for processing large volumes of data but also for visualizing complex relationships between authors, keywords, article topics, and scientific institutions. As a result, researchers can more effectively analyze the structure of scientific knowledge, identify key publications, and forecast future research directions.

A number of recent studies uses specialized bibliometric analysis tools.

Hakkaraki (2024), investigating the evolution of bibliometric analysis from 1974 to 2024 using the Dimensions AI database, focused on citation network analysis, co-authorship, and analysis of scientific institutions, as well as visualization of scientific connections. The analysis results were processed and visualized using *VOSviewer* and Microsoft Excel. Using these methods, the author illustrates trends and the growth of scientific literature in the field of bibliometrics and examines connections between key publications, authors, and organizations. Network analysis shows which publications and authors serve as major nodes in the research network. This helps to identify key players in bibliometrics and understand which scientific studies have the most significant impact on the further development of the discipline.

Raut et al. (2025) presents a comprehensive literature review on healthcare resilience using bibliometric and network analysis methodologies. The authors utilize the *Bibliometrix* package in *R* to analyze a dataset of relevant publications, with special attention to citation patterns, authorship, and keyword co-occurrence.

VOSviewer and *R Studio* were used in (Gandasari et al., 2024) to visualize complex relationships and trends in social network analysis (SNA) literature, and in (Ho & Shekofteh, 2021) to visualize co-authorship and citation networks in highly cited publications on multiple sclerosis.

Phan (2021) conducted a bibliometric analysis to reveal the intellectual structure of social entrepreneurship (SE) research. Using a dataset of 1361 publications from the Web of Science database, the study employed three analytical methods: bibliographic coupling, co-citation, and co-word analyses. These methods were used to identify influential publications, authors, journals, and prevalent keywords within the SE domain. (Rawat & Sood, 2021) also employed *VOSviewer* and *Bibliometrix* software packages for scientometric analysis to evaluate the growth, emerging trends, and global scope of research in big data analytics. (Liu et al., 2015) used *CiteSpace* to visually display the development of research on innovation systems.

Sheng et al. (2024) analyzed 435 highly cited articles using core citation metrics in the field of sports science. The bibliometric indicators of the highly cited articles, including number of publications, authors, countries, institutions, and most cited papers, were presented and analyzed in detail. The Relative Citation Rate (RCR) index was used to identify the most significant articles. This indicator accounts for the number of citations a paper receives over time and enables comparisons of the influence of publications released in different years.

Ganti et al. (2025) used *VOSviewer* and *CiteSpace* for quantitative assessment of scientific activity in medicine. The main methods included trend analysis, structure of scientific collaboration, and intellectual evolution of medical fields. The authors emphasize that such analysis helps to identify research gaps and strategically plan scientific investigations, thereby enhancing their impact and quality.

Alvarez-Peregrina et al. (2021) used citation network analysis to identify key scientific directions and the most influential publications in the field of myopia genetics. *CitNetExplorer* and *CiteSpace* were applied for analysis, and data visualization was conducted using *VOSviewer*. This enabled tracking of publication activity dynamics, identification of main research areas, and determination of interconnections between studies. The citation network analysis showed a significant increase in interest in the topic, especially in recent years, and identified key publications that have had the greatest influence on the development of genetic research on myopia.

Ruiz-Roman et al. (2022) analyzed publications related to COVID-19 and pregnancy using the Web of Science database. *CitNetExplorer* and *VOSviewer* were used to visualize and cluster connections between publications and authors. The study identified the most cited articles and clustered them accordingly.

Furstenau et al. (2021) conducted a bibliometric analysis of the coronavirus literature over the first eight months of 2020 using *SciMAT* software. The main goal of the study was to identify key themes, build thematic networks, and analyze the evolution of scientific areas related to COVID-19.

Other recent articles dedicated to bibliometric analysis in the field of medicine include (Chahrour et al., 2020; Swanson & Santamaria, 2021; Cowhitt & Cutts, 2020; Funada et al., 2023; Rashid & Dudley, 2024; Yeung et al., 2024).

Chen et al. (2022) introduced the Article's Scientific Prestige (ASP) metric for evaluating the impact of scientific articles in the Web of Science. Unlike traditional citation indices, ASP uses eigenvector centrality, taking into account both direct and indirect citations. Analyzing over 63 million articles and 1.45 billion citations across 254 subjects from 1981 to 2020, the authors demonstrated that ASP provides a more objective assessment of scientific contribution, overcoming the limitations of journal rankings and traditional bibliometric indicators.

In (Fonseca et al., 2016), social network analysis (SNA) was used to study collaboration patterns among researchers in the field of healthcare. By constructing co-authorship networks, the study identified key researchers and assessed the structure of scientific collaboration. Data analysis was carried out using specialized software packages such as *Pajek* and *UCINET*.

In (Maltseva & Batagelj, 2024), the authors examine the authorship structure of publications in the field of SNA analysis using citation metrics, particularly direct citations and bibliographic coupling. In their work, the authors employed a variety of research tools

1. *iCite* tool for Relative Citation Ratio (RCR) calculation
2. *VOSviewer* for co-authorship network visualization
3. *PlumX* metrics for assessing altmetric indicators

The study emphasizes the usefulness of combining direct citation analysis with bibliographic coupling to obtain a comprehensive picture of scientific interactions.

Many scientific communities are working during a considerable time, with hundreds of articles published annually, making it necessary to understand the developmental trends of the field over time. Research on network dynamics is applied in various scientific disciplines, including strategic management. Thus, in (Koseoglu, 2016), the evolution of co-authorship networks from 1980 to 2014 is considered, with an emphasis on changes in the citation network over time. Applying methods of bibliometric analysis and social network analytics, the study shows the dynamics of publication activity, structural shifts, and redistribution of influence among scholars. The analysis demonstrates the increasing complexity of the network, the emergence of new research clusters, and the changing roles of key authors. The provided examples of applying various network analysis tools to investigate publications on different topics highlight the relevance of such analysis in bibliometrics.

More than 10 years ago, HSE University (Moscow, Russia) launched development of the system iFORA for intellectual big data analysis. Today it is based on cutting-edge technologies in machine learning and semantic analysis. The system analyses documents in various languages, including English, Russian and Chinese. iFORA contains more than 850 million documents including publications and patents in various domains. The system has been successfully employed for policy development, investment analysis, and validation of research findings. Our colleagues have produced many fundamental studies on forecasting the development of various economic sectors and S&T fields in several countries, using statistical methods and classic models of social network analysis (see, e.g., Gokhberg et al., 2023a; Abdrakhmanova et al., 2023; Gokhberg et al., 2023b; Saritas et al., 2024; Gershman et al., 2025; Sokolov et al., 2025).

Another important aspect of network analysis is term analysis, which is used to identify key concepts and topics in scientific literature. This method is particularly useful for determining major research directions and identifying knowledge gaps.

Cinpolat & Akar (2023) examine the publications from the Web of Science Core Collection, using MeSH (Medical Subject Headings) terms as search criteria. Microsoft Excel, *VOSviewer*, and the *Bibliometrix, Biblioshiny R* package were used for charting and bibliometric analysis. The authors identified key themes, leading countries and research centers, authors, journals, and the publication with the highest co-citation intensity.

As shown by the analysis of publications, the main methods include co-authorship analysis, co-citation, keyword analysis, and thematic clustering. The application of various methodological approaches enables a comprehensive picture of the development of the scientific field, identifies main research trends, and assesses the level of international collaboration. At the same time, network science is also evolving, as reflected in the study by (Molontay & Nagy, 2020), which examines the development of network science over the past two decades.

Mejia et al. (2021) investigate the thematic landscape of bibliometric research by analyzing 20,268 articles containing the terms bibliometrics, scientometrics, or informetrics. Employing methodologies such as citation network analysis and hierarchical semantic relationships, the study uncovers major trends in the field.

2.2 A Survey of the Publications on Bibliometric Analysis of Parkinson's Disease

Bibliometric analysis also plays an important role in the systematic study of scientific publications on the Parkinson's disease (PD). This approach allows for the identification of key trends, research networks, authoritative sources of information, and the most productive areas of scientific work and thematic clusters. In PD research, bibliometric analysis is increasingly used to identify leading scientific centers, analyze interdisciplinary interaction, detect emerging research trends, and assess the contributions of various countries and research teams to the development of this scientific field.

In recent years, network analysis has become one of the key tools in PD research, enabling the study of complex interactions between biological, clinical, and social factors. This methodological approach covers a wide range of applications—from scientific literature analysis to molecular interaction studies—making it indispensable for understanding the multifaceted nature of the disease. Below several examples are presented of the use of network analysis of scientific publications on Parkinson's disease. These publications cover various aspects of PD research. In most studies on bibliometric analysis in the field of PD, the Web of Science and Scopus databases are used as data sources, and the main analytical tools include *VOSviewer*, *CiteSpace*, and *R* packages (*Biblioshiny*, *Bibliometrix*).

In (McBenedict et al., 2024), a comprehensive bibliometric approach was used, including analysis of publication activity by year, identification of leading authors, institutions, and countries, as well as mapping of scientific networks using *VOSviewer*. The authors noted a significant increase in the number of publications since 2000, indicating growing scientific interest in PD. This analysis made it possible to identify leading countries, assess the level of international collaboration, and pinpoint key research centers.

Suresh et al. (2024) presents a comprehensive bibliometric and scientometric analysis of publications on deep brain stimulation (DBS) for Parkinson's disease

(PD), covering the period from 1984 to 2021. The research methodology included co-authorship analysis, keyword cohort analysis, and thematic mapping using *VOSviewer* and *R* Studio (*Bibliometrix* package).

In (Chen et al., 2024a), a bibliometric analysis was conducted on literature concerning the relationship between gut microbiota and PD, published from 2014 to 2023. The relevant literature sources were collected from the Web of Science Core Collection and analyzed using the *Biblioshiny R* package and *VOSviewer*.

In (Cabanillas-Lazo et al., 2022), a bibliometric analysis of scientific output related to gut microbiota and Parkinson's disease was conducted for the period from 2011 to 2020. Using data analysis tools such as *VOSviewer* and *CiteSpace*, the authors identified the most productive and influential authors, affiliations, and journals, as well as the most cited articles.

In (Liao et al., 2024), the authors analyzed publications related to EEG (Electroencephalography) studies in PD from 2004 to 2023, using *CiteSpace* and *VOSviewer* software tools. This study provided valuable insights into major research trends, influential authors, and emerging topics in the field and helped identify leading countries and research centers.

CiteSpace 6.1 R6 software was used in (Pan et al., 2024) to analyze the status, trends, and "hot spots" in the area of speech and language disorders in PD. The analysis focused on annual publication volume, collaboration among countries and institutions, author collaboration, journals, co-citation references, and keywords, aiming to study the current research landscape, hot spots, and frontiers in this area.

Particular attention in bibliometric research is given to the analysis of keywords, as it allows identifying the main trends in scientific publications. In (Chen et al., 2024b), an analysis was conducted on publications regarding the relationship between mitochondrial dysfunction and Parkinson's disease (PD). The methodology employed by the authors included keyword analysis and co-citation analysis using *CiteSpace*, which allowed identifying the main research clusters.

Zhu et al. (2024) analyzed key topics related to sleep disorders in Parkinson's disease, using the method of co-occurrence analysis of keywords. Based on *CiteSpace* and *VOSviewer*, an analysis was conducted from the perspective of trends in the number of annual publications, countries, institutions, authors, journals, and co-citation. As a result, the main research directions were identified, such as circadian rhythms, quality of life, and the impact of sleep on disease progression.

In addition, co-authorship analysis plays an important role in bibliometric research, enabling the assessment of collaboration between scholars and institutions. (Zhang et al., 2023), investigate depression and anxiety in PD patients, co-authorship analysis between countries and institutions is presented. The authors used the Web of Science Core Collection to retrieve publications, and the analysis was carried out using *CiteSpace* and *VOSviewer*. Three promising research directions were identified: functional connectivity, gut microbiota, and inflammation.

Li et al. (2022) presents a review of publications on network visualization of Parkinson's disease research using *CiteSpace* and the Web of Science database. The authors analyzed studies published from 1998 to 2022, paying special attention to co-authorship networks to identify leading researchers, institutions, and countries, as well as their scientific interactions. Key research areas include the study of cognitive impairments in Parkinson's disease and the application of artificial intelligence methods to analyze neuroimaging data.

Recently, publications have emerged regarding the neurological consequences of COVID-19 (Akintunde et al., 2021). A systematic review of studies has shown that there may be a link between COVID-19 infection and mental health, particularly the development of Parkinson's disease (Tsai et al., 2020; Fearon & Fasano, 2021; Beauchamp et al., 2020; Brundin et al., 2020). This factor may also contribute to the increase in the number of people with Parkinson's disease.

2.3 Network Analysis in Bibliometrics

The development of network science methods, such as graph theory, statistics, and informatics, has also impacted the field of bibliometrics. The concept of centrality, using its various numerical characteristics—centrality indices—is also employed in network bibliometric analysis. The use of centrality metrics helps to understand which nodes in the network are most influential, how knowledge spreads within the academic community, and what factors may limit access to scientific information. These approaches are applied in (Duggan et al., 2024) concerning publication activity in the field of anesthesiology. The study uses social network analysis methods to identify patterns of interaction between authors and research institutions, as well as to identify significant imbalances in the influence levels of different network participants. Centrality analysis was applied to assess the influence of authors and organizations in the network. The article uses several centrality metrics, such as Degree, Eigenvector, and Betweenness, to identify the most influential authors and institutions. Using metrics such as the Gini coefficient, the authors also assessed the degree of inequality in the distribution of influence among nodes. The Gini coefficient indicated that there is significant inequality both among authors and among institutions, which may affect the distribution of knowledge and opportunities for collaboration. Thus, network analysis in this context makes it possible not only to see the current structure of the scientific network but also to identify areas for improving collaboration, more equitable information distribution, and inclusion of diverse perspectives in future research.

A variety of quantitative centrality measures (indices) have been developed for the study of social networks. However, classical indices do not always account for the structure of the network and the individual characteristics of specific actors, as well as the intensity of direct and indirect, particularly intragroup, interactions within the system. Therefore, alternative centrality measures have been proposed, in particular, short-range interactions centrality (SRIC) and long-range interactions centrality (LRIC), which make it possible to take these specific features into account. The SRIC index considers both the direct influence of one vertex in the network on another and the influence mediated through other vertices directly connected to the vertex under consideration. The LRIC index is an improvement over SRIC; its distinctive feature lies in the additional consideration of all possible indirect influence channels from one vertex to another. The LRIC index has several modifications (Sum, Max, MaxMin, MaxT, MultT, Simul), each of which calculates the intensity of influence of each individual channel in its own way, as well as their cumulative influence. These indices were proposed in (Aleskerov et al., 2014) (SRIC was referred to as KBI there). These indices were used in (Aleskerov et al., 2016) to identify the most important international and Russian economic journals in networks that arise from cross-citation analysis. A detailed description of these indices is given in (Aleskerov et al., 2022).

In (Aleskerov et al., 2017), models of network analysis of patents and publications on various aspects of Parkinson's disease using SRIC and LRIC were presented. The results obtained made it possible to identify links between research clusters, rank their significance, and track changes in the attention of scientific groups—for example, to previously unknown patents and developments.

In (Aleskerov et al., 2018), methods of network and semantic analysis were used to study publications on Parkinson's disease from the Web of Science database. The authors applied centrality indices (SRIC and LRIC) to identify influential articles and used the LDA (Latent Dirichlet Allocation) algorithm for thematic analysis.

New indices—BI (Bundle Index, group influence index) and PI (Pivotal Index, pivotal node influence index)—were proposed in (Aleskerov & Yakuba, 2020). These indices were used in (Aleskerov et al., 2023) and (Aleskerov et al., 2024) for network analysis of affiliations and publications related to Parkinson's disease (PD).

References

Abdrakhmanova, G., Arundel, A., De Fuentes, C., de Jong, P. J., De Marchi, V., Gokhberg, L., Roud, V., Streltsova, Y., & Vishnevskiy, K. (2023). *Handbook of innovation indicators and measurement*, 2nd edition. In F. Gault & A. Arundel (Eds.), *E*. Edward Elgar Publishing.

Akintunde, T. Y., Musa, T. H., Musa, H. H., Musa, I. H., Chen, S., Ibrahim, E., Tassang, A., & Helmy, M. (2021). Bibliometric analysis of global scientific literature on effects of COVID-19 pandemic on mental health. *Asian Journal of Psychiatry, 63*, 102753. https://doi.org/10.1016/j.ajp.2021.102753

Aleskerov, F., Andrievskaya, I., & Permjakova, E. (2014). Key borrowers detected by the intensities of their short-range interactions. *HSE University Basic Research Program Working Paper*, Series Financial Economics, WR BRP, 33/FE/2014, ssrn.com/abstract=2479272

Aleskerov, F., Buldyaev, A., Khutorskaya, O., & Yamilov, A. (2017). Parkinson's disease: network analysis of patents and publications' activity. *Proceedings of the 11th IEEE International Conference on Application of Information and Communication Technologies* (AICT2017, Moscow) 430–432. ISBN/ISSN: 978–1–5386-0500-4. doi:https://doi.org/10.1109/ICAICT.2017.8686912.

Aleskerov, F., Khutorskaya, O., Buldyaev, A., & Yamilov, A. (2018). Parkinson's disease: Network analysis of publications' impact. *In 7th international conference on computers communications and control* (ICCCC 2018, Oradea, Romania), 82–85. IEEE. ISBN/ISSN: 978–1–5386-1934-6.

Aleskerov, F., Khutorskaya, O., Stepochkina, A., Yakuba, V., & Zinoveva, K. (2023). Network analysis of publications on studies of Parkinson disease. Procedia Computer Science, 219, 1380–1387. doi:https://doi.org/10.1016/j.procs.2023.01.426 www.sciencedirect.com/science/article/pii/S1877050923004350.

Aleskerov, F., Khutorskaya, O., Stepochkina, A., Yakuba, V., & Zinoveva, K. (2024). Affiliations based bibliometric analysis of publications on Parkinson's disease. *J Computational Management Science, 21*(13), 1–14. https://doi.org/10.1007/s10287-023-00495-7

Aleskerov, F., Pislyakov, V., & Subochev, A. (2016). Ranking journals using social choice theory methods: A novel approach in bibliometrics. in: *Proceedings of the 21 International Conference on Science and Technology Indicators*. Ed. by I. Ràfols, J. Molas-Gallart, E. Castro-Martínez, R. Woolley. Valencia: Editorial Universitat Politècnica de València, 16, 1266–1273. doi: https://doi.org/10.4995/STI2016.2016.4543.

Aleskerov, F., Shvydun, S., & Meshcheryakova, N. (2022). New centrality measures in networks: How to take into account the parameters of the nodes and group influence of nodes to nodes. *CRC Press*. https://doi.org/10.1201/9781003203421

Aleskerov, F., & Yakuba, V. (2020). Matrix-vector approach to construct generalized centrality indices in networks. *WP7/2020/01*. Moscow: Higher School of Economics Publ. House, SSRN. ssrn.com/abstract=3597948

Alvarez-Peregrina, C., Martinez-Perez, C., Villa-Collar, C., & Sánchez-Tena, M. (2021). A bibliometric and citation network analysis of myopia genetics. *Genes, 12*(3), 447. https://doi.org/10.3390/genes12030447

Aria, M., & Cuccurullo, C. (2017). Bibliometrix: An R-tool for comprehensive science mapping analysis. *Journal of Informetrics, 11*(4), 959–975. https://doi.org/10.1016/j.joi.2017.08.007

Beauchamp, L. C., Finkelstein, D. I., Bush, A. I., Evans, A. F., & Barnham, K. J. (2020). Parkinsonism as a third wave of the COVID-19 pandemic? *Journal of Parkinson's Disease, 10*(4), 1343–1353. https://doi.org/10.3233/JPD-202211

Brundin, P., Nath, A., & Beckman, J. D. (2020). Is COVID-19 a perfect storm for Parkinson's disease? *J Trends in Neurosciences, 43*(12), 931–933. https://doi.org/10.1016/j.tins.2020.10.009

Cabanillas-Lazo, M., Quispe-Vicuna, C., Barja-Ore, J., Fernandez-Giusti, A., Munive-Degregori, A., Retamozo-Siancas, Y., Guerrero, M. E., & Mayta-Tovalino, F. (2022). A 10-year bibliometric analysis of global research on gut microbiota and Parkinson's disease: Characteristics, impact, and trends. *Hindawi. BioMed Research International, 4144781, ISSN*, 2314–6133. https://doi.org/10.1155/2022/4144781

Chahrour, M., Assi, S., Bejjani, M., Nasrallah, A. A., Salhab, H., Fares, M., & Khachfe, H. H. (2020). A bibliometric analysis of COVID-19 research activity: A call for increased output. *Cureus, 12*(3), e7357. https://doi.org/10.7759/cureus.7357

Chen, Y., Koch, T., Zakiyeva, N., Liu, K., Xu, Z., Chen, Ch-h., Nakano, J., & Honda, K. (2022). Article's Scientific Prestige: measuring the impact of individual articles in the Web of Science. *arXiv*:2202.08695 [cs.DL]. doi: 10.48550/arXiv.2202.08695.

Chen, Q., Sun, T., & Wang, T. (2023). Network centrality, support organizations, exploratory innovation: Empirical analysis of China's integrated circuit industry. *J Heliyon, 9*(7), e17709. https://doi.org/10.1016/j.heliyon.2023.e17709

Chen, L., Chen, J., Wu, M., Yan, P., & Zhou, X. (2024a). Analyzing the bibliometrics of brain-gut axis and Parkinson's disease. *Frontiers in Neurology, 15*, 1343303. https://doi.org/10.3389/fneur.2024.1343303

Chen, Y.-J., Xie, M.-R., Zhou, S.-Q., & Liu, F. (2024b). Mapping the research of mitochondria and Parkinson's disease: A bibliometric analysis. *Frontiers in Neurology, 15*, 1413762. https://doi.org/10.3389/fneur.2024.1413762.

Cinpolat, H. Y., & Akar, A. (2023). A visual and bibliometric analysis of Parkinson's disease biomarker research. *Ann Med Res, 30*(10), 1359–1364. https://doi.org/10.5455/annalsmedres.2023.09.253

Cowhitt, T., & Cutts, A. (2020). Using network analysis to compare bibliographic database journal coverage. *Journal of Electronic Resources Librarianship, 32*(3), 195–210. https://doi.org/10.1080/1941126X.2020.1790952

Duggan, E. W., Atwood, G. S., Sanford, J. A., Tsai, M. H., Egbaria, J. K., Carmichael-Tanaka, N., & Outland, N. B. (2024). Using bibliometric data to define and understand publishing network equity in anesthesiology. *Anesthesia & Analgesia, 139*(5), 944–954. https://doi.org/10.1213/ANE.0000000000006877

Fearon, C., & Fasano, A. (2021). Parkinson's disease and the COVID-19 pandemic. *Journal of Parkinson's Disease, 11*(2), 431–444. https://doi.org/10.3233/JPD-202320

Fonseca, B. D., Sampaio, R. B., Fonseca, M. V., & Zicker, F. (2016). Co-authorship network analysis in health research: Method and potential use. *Health Research Policy and Systems., 14*, 34. https://doi.org/10.1186/s12961-016-0104-5

Funada, S., Yoshioka, T., Luo, Y., Iwama, T., Mori, V., Yamada, N., Yoshida, H., Katanoda, K., & Furukawa, T. A. (2023). Global trends in highly cited studies in COVID-19 research. *JAMA Network Open, 6*(9), e2332802. https://doi.org/10.1001/jamanetworkopen.2023.32802

Furstenau, L. B., Rabaioli, B., Sott, M. K., Cossul, D., Bender, M. S., De Mattos, E. M. J., Barcellos, F. N., Severo, P. P., Dohan, M. S., & Bragazzi, N. L. (2021). A bibliometric network analysis of coronavirus during the first eight months of COVID-19 in 2020. *International Journal of Environmental Research and Public Health, 18*(3), 952. https://doi.org/10.3390/ijerph18030952

Gandasari, D., Tjahjana, D., Dwidienawati, D., & Sugiarto, M. (2024). Bibliometric and visualized analysis of social network analysis research on Scopus databases and VOSviewer. *J Cogent Business & Management, 11*(1), 2376899. https://doi.org/10.1080/23311975.2024.2376899

Ganti, L., Persaud, N. A., & Stead, T. S. (2025). Bibliometric analysis methods for the medical literature. *Academic Medicine & Surgery., 129134* .https://doi.org/10.62186/001c.129134

Gershman, M., Gokhberg, L., & Kuznetsova, T. (2025). Science, technology, and innovation in Russia: Development challenges and opportunities. In: Zhao X, Sokolov A, Cassiolato J E (eds) The innovation competitiveness of BRICS countries. Science, technology and innovation studies. Springer, Singapore, 207–225. ISSN2570–1509. doi: https://doi.org/10.1007/978-981-97-8001-3_10.

Gokhberg, L., Abdrakhmanova, G., Streltsova, E., & Vishnevskiy, K. (2023a). Measuring the digital transformation. In: *Handbook of innovation indicators and measurement*, 2nd edition. Ed. by F. Gault, A. Arundel, E. Kraemer-Mbula. Edward Elgar Publishing, 12, 221–239.

Gokhberg, L., Kuznetsova, T., & Kotsemir, M. N. (2023b). From the Soviet Union to The Russian Federation: Publication activity dynamics along the evolution of national science policies. *Scientometrics, 128*(11), 6195–6246. https://doi.org/10.1007/s11192-023-04838-8

HabibAgahi, M. R., Kermani, M. A., & Maghsoudi, M. (2022). On the co-authorship network analysis in the process mining research community: A social network analysis perspective. *Expert Systems with Applications, 206*, 117853. https://doi.org/10.1016/j.eswa.2022.117853

Hakkaraki, V. P. (2024). Exploring the evolution of bibliometric analysis: A comprehensive study of scientific publications from 1974 to 2024 using the dimensions AI database. *Asian Journal of Information Science and Technology, 14*(1), 24–31. https://doi.org/10.70112/ajist-2024.14.1.3878

Ho, Y. S., & Shekofteh, M. (2021). Performance of highly cited multiple sclerosis publications in the science citation index expanded: A scientometric analysis. *Multiple Sclerosis and Related Disorders, 54*, 103112. https://doi.org/10.1016/j.msard.2021.103112

Koseoglu, M. A. (2016). Growth and structure of authorship and co-authorship network in the strategic management realm: Evidence from the strategic management journal. *BRQ Business Research Quarterly, 19*(3), 153–170. https://doi.org/10.1016/j.brq.2016.02.001

Li, X. L., Gao, R. X., Zhang, Q., Li, A., Cai, L.-N., Zhao, W.-W., Gao, S.-L., Wang, Y., & Yue, J. (2022). A bibliometric analysis of neuroimaging biomarkers in Parkinson disease based on web of science. *Medicine (Baltimore), 101*(33), e30079. https://doi.org/10.1097/MD.0000000000030079

Liao, X. Y., Gao, Y. X., Qian, T. T., Zhou, L.-H., Li, L.-Q., Gong, Y., & Ye, T. F. (2024). Bibliometric analysis of electroencephalogram research in Parkinson's disease from 2004 to 2023. *Front. Neurosci., Sec. Neurodegeneration, 18*, 1433583. https://doi.org/10.3389/fnins.2024.1433583

Liu, Z., Yin, Y., Liu, W., & Dunford, M. (2015). Visualizing the intellectual structure and evolution of innovation systems research: A bibliometric analysis. *Scientometrics, 103*, 135–158. https://doi.org/10.1007/s11192-014-1517-y

Maltseva, D., & Batagelj, V. (2024). Citation and bibliographic coupling between authors in the field of social network analysis. *Journal of Data and Information Science, 9*(4), 110–154. https://doi.org/10.2478/jdis-2024-0028

McBenedict, B., Hauwanga, W. N., Ienaco, G., Dulci, P., Syeda, S. K., Jonatha, M. L., Barakat, B. O., Asaju, F., Sujood, A., Shivadeva, S., Phoh, W. C., & Bruno, L. P. (2024). Parkinson's disease treatment: A bibliometric analysis. *Cureus, 16*(9), e69613. https://doi.org/10.7759/cureus.69613

Mejia, C., Wu, M., Zhang, Y., & Kajikawa, Y. (2021). Exploring topics in bibliometric research through citation networks and semantic analysis. *Front. Res. Metr. Anal., 6*, 742311. https://doi.org/10.3389/frma.2021.742311

Minati, G. (2024). Linked links—A research project: The multiple superimposed soft networks as network profiles. *Systems, 12*(8), 303. https://doi.org/10.3390/systems12080303

Molontay, R., & Nagy, M. (2020). Twenty years of network science: a bibliographic and co-authorship network analysis. *arXiv*:2001.09006 [physics.soc-ph]. doi:10.48550/arXiv.2001.09006.

Newman, M. E. (2018). *Networks* (Second ed.). Oxford University Press.

Pan, X., Liang, B., & Cao, T. (2024). A bibliometric analysis of speech and language impairments in Parkinson's disease based on web of science. *Front. Psychol., Sec. Psychology of Aging, 15*, 1374924. https://doi.org/10.3389/fpsyg.2024.1374924

Phan, T. L. (2021). Mapping the social entrepreneurship research: Bibliographic coupling, co-citation and co-word analyses. *Cogent Business & Management, 8*(1), 1896885. https://doi.org/10.1080/23311975.2021.1896885

Rashid, F., & Dudley, J. (2024). Bibliometric analysis of the 100 most-cited articles on the methods of shade-matching in dentistry. *Clinical and Experimental Dental Research, 10*(6), e70037. https://doi.org/10.1002/cre2.70037

Raut, S., Hossain, N., & Fazio, S. (2025). Healthcare resilience: Bibliometric and network analysis. *Enterprise Information Systems, Art., No. 2464752.* https://doi.org/10.1080/17517575.2025.2464752

Rawat, K. S., & Sood, S. K. (2021). Emerging trends and global scope of big data analytics: A scientometric analysis. *Quality and Quantity, 55*, 1371–1396. https://doi.org/10.1007/s11135-020-01061-y

Ruiz-Roman, R., Martinez-Perez, C., Gil Prados, I., Cristóbal, I., & Sánchez-Tena, M. Á. (2022). COVID-19 and pregnancy: Citation network analysis and evidence synthesis. *JMIR Pediatr Parent, 5*(1), e29189. https://doi.org/10.2196/29189

Saritas, O., Veselitskaya, N., & Sokolov, A. (2024). Foresight methodology and commonly used methods in the new generation foresight. In O. Saritas, A. Sokolov, & M. Cele (Eds.), *In: 21st century foresight: Shaping the future for sustainable social, economic and environmental development in South Africa* (Vol. 4, pp. 35–53). Springer. https://doi.org/10.1007/978-3-031-69557-5_4

Sheng, Y., Huiting, Z., Qiang, Z., & Chenhui, L. (2024). A corpus-based bibliometric study of highly cited papers in sport sciences. *SAGE Open, 14*(1). https://doi.org/10.1177/21582440231225856

Sokolov, A., Shashnov, S., & Kotsemir, M. (2025). Intra-BRICS research collaboration: Status and prospects. In: Zhao X, Sokolov A, Cassiolato J.E. (eds) The innovation competitiveness of BRICS countries. *Science, Technology and Innovation Studies*, Springer, 3, 41–65. ISBN 978-981-97-8000-6. doi: https://doi.org/10.1007/978-981-97-8001-3_2.

Suresh, V., Dave, T., Ghosh, S., Jena, R., & Sanker, V. (2024). Deep brain stimulation in Parkinson's disease: A scientometric and bibliometric analysis, trends, and research hotspots. *Medicine, 103*(20), e38152. https://doi.org/10.1097/MD.0000000000038152

Swanson, D., & Santamaria, L. (2021). Pandemic supply chain research: A structured literature review and bibliometric network analysis. *Logistics, 5*(1), 7. https://doi.org/10.3390/logistics5010007

Tsai, S. T., Lu, M. K., San, S., & Tsai, C. H. (2020). The neurologic manifestations of coronavirus disease 2019 pandemic: A systemic review. *Frontiers in Neurology, 11*, 498. https://doi.org/10.3389/fneur.2020.00498

Yeung, A. W. K., Litvinova, O., Bragazzi, N. L., et al. (2024). Digital health and mobile health: A bibliometric analysis of the 100 most cited papers and their contributing authors. *Explor Digit Health Technol, 2*, 86–100. https://doi.org/10.37349/edht.2024.00013

Zhang, T., Yang, R., Pan, J., & Huang, S. (2023). Parkinson's disease related depression and anxiety: A 22-year bibliometric analysis (2000-2022). *Neuropsychiatric Disease and Treatment, 19*, 1477–1489. https://doi.org/10.2147/NDT.S403002

Zhu, L., Zhu, P., Wang, J., Yan, K., Zhao, S., Jiang, Y., & Zhang, H. (2024). A bibliometric and visual analysis of Parkinson's disease sleep disorders: Articles from 2008 to 2023. *Frontiers in Psychiatry, 15*, 1468568. https://doi.org/10.3389/fpsyt.2024.1468568

Chapter 3
Data Description and Network Analysis Models

3.1 Data Formation Algorithm

The Microsoft Academic scientific publication search system was used (date of access to the database: 20.11.2021) to access data—an open and publicly accessible search engine for scientific publications developed by Microsoft Research in 2006 (Sinha et al., 2015; Wang et al., 2019). The system indexes approximately 250 million publications, including scientific articles, books and individual chapters, patents, and conference proceedings.

To construct the citation network and analyze centrality, articles were selected that met the following query: the words "parkinson" and "disease" are present in the title/abstract for the years 2015–2021. This time frame was chosen to trace the main research interests and trends in their evolution over recent years. It should be noted that the 2021 data are presented as of the access date to the database, meaning that not all articles written in that year may have been published at the time of data download. The data download and subsequent computations were performed using the high-level programming language Python within the JupyterLab development environment. This environment provides a convenient interface for creating data analysis documents. Data download was implemented using Python libraries urllib, requests, and http. These libraries enable queries to the Microsoft Academic database API and return responses in JSON format.

All data were compiled into a single CSV table. Abstracts and titles of articles were normalized, i.e., all words were reduced to their base form (the main immutable form of the word from which all its grammatical variants are derived, e.g., for verbs—this is the infinitive). Each article had more than 30 attributes. The following 17 attributes were used in this study

© The Editor(s) (if applicable) and The Author(s), under exclusive license to Springer Nature Switzerland AG 2026

F. Aleskerov et al., *Bibliometric Analysis by Network Models*, Contemporary Systems Thinking, https://doi.org/10.1007/978-3-032-09171-0_3

- Id—unique publication identifier. Example: Id 2754967293
- Ti—normalized title. A title in which all words are written in lowercase letters. Example: "global regional and national burden of neurological disorders 1990–2016, a systematic analysis for the global burden of disease study 2016"
- W—unique normalized words in the title. Example: ['global', 'regional', 'national', 'burden', 'neurological', 'disorders', 'systematic', 'analysis']
- AW—unique normalized words in the abstract. Example: ['summary', 'background', 'comparable', 'data', 'global', 'country', 'specific']
- Y—year of publication. Example: 2017
- CC—citation count. Example: 1535.
- RId—list of IDs of the articles referenced by the publication. Example: [RId 2159011576, RId 2751884637, RId 2125065061]
- DN—title. Example: "Global, regional, and national burden of neurological disorders, 1990–2016: A systematic analysis for the Global Burden of Disease Study 2016"
- DOI—digital object identifier. Example: 10.1016/S1474-4422(18)30499-X
- IA—annotation in Inverted Index format. Example: {'summary': [1], 'background': [5], 'comparable': [10], 'data': [15, 20], 'global': [25], 'country': [30], 'specific': [35]}. In this example, the numbers indicate the positions of the respective words in the abstract, e.g., 'summary': [1] means that the word 'summary' occurs in the first position of the abstract, while 'data': [15, 20] means that the word 'data' appears in the 15th and 20th positions.
- AA.AfId—author's affiliation ID. Example: AA.AfId 39854758
- AA.AfN, AA.DAfN—name of the author's affiliation. Example: 'AfN': 'auckland university of technology', 'DAfN': 'Auckland University of Technology'
- AA.AuId—author ID. Example: AA.AuId 130753112
- AA.AuN, AA.DAuN—author's name. Example: 'AuN': 'valery l feigin', 'DAuN': 'Valery L Feigin'
- J.JId—journal ID. Example: J.JId 70053155
- J.JN—journal name. Example: "Lancet Neurology"
- D—publication date in the format YYYY-MM-DD. Example: 2017-11-01.

For simplification, some fields such as authors and their affiliations, as well as word positions in the abstract, were shortened and simplified to preserve the basic data structure without loss of informativeness.

Figure 3.1 presents a fragment of the downloaded dataset with some attributes.

A total of 70,119 articles were downloaded. To improve the quality of network construction and subsequent analysis, additional processing was carried out. This stage included preliminary data analysis, processing of abstracts of the articles, removal of missing values and errors, graph preparation (converting the data into the "source vertex – target vertex" structure), and semantic analysis. Isolated vertices were not included in the network. For this stage, Python libraries were used: numpy,

AA	AW	CC	D	DN	DOI
[{'DAuN': 'Ronald B. Postuma', 'AuN': 'ronald b postuma', 'AuId': 2047227543, 'DAfN': 'Montreal General Hospital', ...	['document', 'movement', 'disorder', 'society', ...	2014	01.10.2015	MDS clinical diagnostic criteria for Parkinson's disease	10.1002/MDS.26424
[{'DAuN': 'B. K. Pedersen', 'AuN': 'bente klarlund pedersen', 'AuId': 2127745026, 'DAfN': 'University of Copenhagen' ...	['review', 'reader', 'date', 'evidence', 'basis', 'prescribing', ...	1241	01.12.2015	Exercise as medicine – evidence for prescribing exercise as ...	10.1111/SMS.12581
[{'DAuN': 'Timothy R. Sampson', 'AuN': 'timothy r sampson', 'AuId': 2112352720, 'DAfN': 'California Institute of Technology', ...	['intestinal', 'microbiota', 'influence',...	1347	01.12.2016	Gut Microbiota Regulate Motor Deficits and ...	10.1016/J.CELL.2016.11.018

Fig. 3.1 Fragment of the downloaded raw data array

pandas, matplotlib. The next step involved graph construction, network analysis, and community detection.

Analysis of the downloaded articles also revealed that the set contained 115 articles related to Danon disease. One of the symptoms of this disease is Wolff-Parkinson-White syndrome, a congenital heart defect. This syndrome is not related to Parkinson's disease; therefore, articles with the words 'wolf' and 'parkinson', 'wpw', 'wpws', 'danon' in the title or abstract were removed from the network.

For the correct construction of the citation graph of publications, the DOI of the works was used, since this field is present in the article descriptions and the identification numbers are indicated in the citation field. However, detailed data analysis showed that the DOI was not specified for all works—10,681 publications lacked DOI information. Therefore, the number of publications used in graph construction decreased to 59,438. Of these, 7315 had no abstracts, and 52,132 had them. Journal information was available for 45,940 articles.

Further analysis showed that a number of articles referenced publications that were released after the citing article, meaning an article from the past cited a future one. A total of 4184 such edges were removed. For further analysis, only articles with DOI, annotation, and journal information were used to enable result interpretation.

After all filtering steps, the network consisted of 39,811 articles, from which 3361 terms were obtained for further processing in the analysis of co-occurrence of terms. The author network comprised 27,551 authors and 3029 affiliations. A total of 3292 journals were analyzed.

3.2 Network Construction from the Obtained Data

Let $G = (V, E)$ be a graph, where s is a non-empty set of vertices, $E \subseteq V \times V$ is a set of edges, $A = [a_{ij}]$ is the adjacency matrix ($a_{ij} = 1$ if there is an edge from vertex i to vertex j, otherwise $a_{ij} = 0$), $W = [w_{ij}]$ is the weight matrix (w_{ij} is the weight of the edge from vertex i to vertex j).

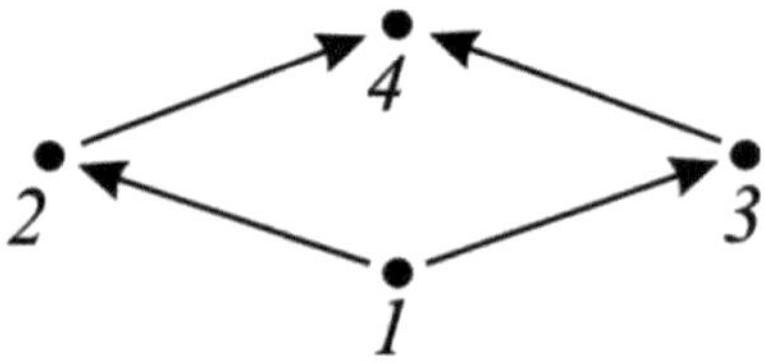

Fig. 3.2 Example of a directed graph with 4 vertices

A graph is called directed if each edge has a direction, given by an ordered pair of vertices (i, j). The existence of an edge (i, j) indicates a connection from i to j.

A graph is called unweighted if $\forall i_1, i_2, j_1, j_2 \in V$: $(i_1, j_1), (i_2, j_2) \in E \Rightarrow w_{i_1 j_1} = w_{i_2 j_2}$ for all edges, i.e., all edge weights are equal. Otherwise, a graph is called weighted.

To calculate graph characteristics and centrality indices, it is necessary to represent the graph as an adjacency matrix $A = [a_{ij}]_{n \times n}$, where $a_{ij} = 1$ if $(i, j) \in E$, and $a_{ij} = 0$ otherwise, and a weighted adjacency matrix $W = [w_{ij}]_{n \times n}$), where w_{ij} is the weight of the edge from i to j.

Let us provide an example for the above definition using a graph with 4 vertices and several directed edges, along with its adjacency matrix.

Let there be a directed, unweighted graph $G=(V, E)$, where,

$V = \{v_1, v_2, v_3, v_4\}$—set of vertices,

$V = \{(v_1, v_2), (v_1, v_3), (v_2, v_4), (v_3, v_4)\}$—set of directed edges.

Figure 3.2 presents an example of a directed graph with 4 vertices.

The adjacency matrix A is

$$\begin{matrix} 0 & 1 & 1 & 0 \\ 0 & 0 & 0 & 1 \\ 0 & 0 & 0 & 1 \\ 0 & 0 & 0 & 0 \end{matrix}$$

The following types of graphs are considered in the study.

Publication citation graph. The purpose of constructing this graph is to determine the influence of vertices. The vertices of the graph are publications, the edges represent citations. The type of graph is directed. We analyze a graph containing 39,811 vertices and 310,829 edges.

Citation analysis allows the identification of key research directions and determination of trends in their development.

Journal citation graph. Information for each publication contains the journal in which it was published, which allows the construction of a network of interactions among scientific journals on a specific topic. The graph in this case is constructed as follows: vertices represent journals, and the edge weight is the number of references from one journal to another over a certain period. The type of graph is weighted and directed (edges are assigned weights—the number of references). Unlike publications, journals may cite themselves. If journals cite themselves, loops are formed in the graph (a vertex is connected to itself). The constructed journal citation graph contained 3292 vertices and 152,203 edges.

Journal citation analysis makes it possible to identify the most influential journals, determine the journal's rank in the research field, its productivity, and scientific popularity.

Author citation graph. Each publication also contains information about the authors. For each article, it was decided to select the first author from the list, as in medical articles the first author is usually the leading author or the project leader. Thus, the vertices of the graph are the first authors of the articles.

An edge corresponds to the citation of an article by author A of an article by author B. The edge weight is the number of citations for a given year. The constructed author citation graph contained 27,551 author vertices and 271,623 edges.

The analysis of the author publication activity graph is mainly carried out to identify the most significant authors.

Affiliation citation graph. At the data processing stage, the affiliations of researchers were identified for each article by processing the relevant fields using the Python programming language. As a result, a list of affiliations for the authors of each article was obtained. Two graphs were constructed—a network based on the first author's affiliation and based on all authors' affiliations. The network based on the first author's affiliation contained 3029 organization vertices and 128,602 edges. The analysis of such a network determines the significance of scientific teams and research centers in a given research field.

To analyze the organizational collaboration network, a graph was constructed in which the vertices represented the affiliations of all authors, and edges were drawn between them if authors from these affiliations had co-authored publications. The organizational collaboration network is an undirected graph with 4697 vertices and 85,900 edges.

Term co-occurrence graph. The research topic can often be determined by the terminology used in publications. Therefore, term analysis provides a broad understanding of the development of various research directions and helps to identify future prospects. From the abstracts of the downloaded publications, we identified six sets of terms that define different research directions on Parkinson's disease. These sets include from 20 to 147 terms.

The following six research directions were identified

1. Biochemistry, cell, genetics (BCG)
2. Medicines
3. Symptoms
4. Central nervous system (CNS)
5. Diagnostics
6. Research methods

For each of the six research directions, networks of term co-occurrence in articles were constructed, taking into account article centrality. The networks were built as follows.

An article abstracts should contain at least a pair of terms from the same topic. There may be several such pairs. These terms are the vertices of the network. An arc is present between the terms if a pair of terms appears together in article abstracts. The weight of the undirected arc between terms (network vertices) is calculated as the sum of centralities of all articles in which this pair of terms is simultaneously mentioned.

Thus, for each research direction, three networks are considered, with weights based on In-degree, BI, and PI centrality indices of articles in abstracts of which the terms appear (for a description of centrality indices, see 3.3). For each network, the centrality of terms is calculated. For the network based on article centrality index In-degree, term centrality is calculated using the In-degree index; for the network based on BI index, term centrality is calculated using BI; and for the network based on PI article centrality, term centrality is calculated using the PI index accordingly. For the BI and PI indices, the quota is taken as 1% with groups size limitation k =5.

Table 3.1 presents the main indicators of term networks (In-degree, BI, and PI) for the 6 research directions on Parkinson's disease.

Table 3.1 Main indicators of term networks

Index name	BCG	Medicines	Symptoms	CNS	Diagnos	Methods
	1	2	3	4	5	6
Number of vertices	139	147	55	39	23	20
Number of edges	6269	1008	830	568	172	136
Maximum edge weight (in-degree index)	0.183	0.0102	0.0473	0.0403	0.018	0.0406
Average edge weight (in-degree index)	0.0011	0.0002	0.0017	0.0015	0.0012	0.0012
Maximum edge weight (BI index, q = 1%)	0.0098	0.0207	0.0988	0.0806	0.0258	0.0209
Average edge weight (BI index, q = 1%)	0.0012	0.0005	0.0024	0.0011	0.0011	0.0007
Maximum edge weight (PI index, q = 1%)	0.3643	0.0331	0.2105	0.1569	0.0350	0.0329
Average edge weight (PI index, q = 1%)	0.0012	0.0011	0.0038	0.0007	0.0008	0.0003

3.3 Theoretical Description of Network Analysis Methods

3.3.1 Centrality Indices—Classical and New

Various methods can be used to identify the most significant nodes (also called vertices) in a network. The characteristic "centrality" allows determining the degree of importance of a graph node based on its position. Centrality in a graph is a vector that assigns a certain number (index) to each node of the graph, indicating its importance. By calculating centrality indices and ranking them, the most influential nodes in a graph can be identified and analyzed. There is a large number of different indices that can be calculated for a network depending on the task. A detailed description of these indices is given in (Newman, 2010).

A description of the networks we constructed was given above. For these networks, we used the following classical indices. In the task of identifying the most influential nodes in a network, the in-degree centrality index is used, as it shows the number of links to a node.

Weighted in-degree centrality—the sum of the weights of incoming edges, i.e., the number of citations of the node if the graph is a citation network

$$In - degree(i) = \sum_{j=1}^{n} w_{ij}$$

where i—number of the considered node, w_{ij}—weight of edge (i, j), n—number of nodes in the graph.

Degree centrality does not take into account any other parameters of the node or the network as a whole. For example, a node with a small sum of incoming edges, but referenced by several popular nodes, should also have high centrality. To take into account such connections, the following centrality indices were introduced.

Eigenvector centrality was described in (Bonacich, 1972) and can be evaluated by the following formula

$$x_i = \frac{1}{\lambda_1} \sum_{j}^{n} A_{ij} x_j$$

where λ_1—the largest eigenvalue of the adjacency matrix A, x—eigenvector, i.e., in vector form, this formula is written as $\lambda x = Ax$.

Eigenvector centrality determines the importance of a node by considering the importance of its neighbors. However, in some cases, it is also worth considering the number of outgoing edges of nodes. For example, if there is a node with high centrality that refers to many other nodes, their centrality should not increase as much as if the important node referred to only one node. In terms of a journal citation network, if there is a significant journal in which various articles are published that cite a large number of other journals, their centrality should not increase as much as if those articles cited only one journal. It is necessary to consider the fact that in a large and popular journal, there are always more references than in less significant journals.

To also take into account the out-degree of a node, the PageRank Index (Brin & Page, 1998) was introduced. Essentially, this index is a variation of eigenvector centrality, but the centrality value of the nodes neighboring the given node is divided by the number of their outgoing edges, i.e.

$$x_i = \alpha \sum_{j}^{n} A_{ij} \frac{x_j}{k_j^{out}} + \beta$$

Constants α and β are used to avoid zero index values.

Another centrality index is Betweenness (Freeman 1977). This index determines the proportion of paths between two nodes that pass through the given node

$$x_i = \sum_{kj}^{n} \frac{\sigma_{kj}^{i}}{\sigma_{kj}}$$

where σ_{kj}—number of shortest paths between nodes k and j, σ_{kj}^{i}—number of shortest paths between nodes k and j that pass through node i.

Ranking by Betweenness centrality index usually differs significantly from ranking by other centrality indices. It makes sense to use such a centrality index to identify nodes that serve as connecting elements in a network. In terms of bibliometric analysis, a node with a high Betweenness index may represent an important interdisciplinary study. It is important to note that each of the described classical indices was calculated for the networks and normalized so that the sum of all values equals to 1.

The described indices are classical and are used in most works on network analysis. A serious drawback of these centrality indices is the lack of consideration for node parameters and ways to account for group influence. To take into account these and other node parameters, new centrality indices were introduced (Aleskerov & Yakuba, 2020). These are the BI (Bundle Index)—the group influence index, and the PI (Pivotal Index)—the pivotal node influence index. To describe these indices, certain definitions must be introduced.

For each node, a quota q_i is defined. As a rule, the quota is calculated as a certain percentage of the sum of the weights of incoming edges. For each node, the value of the quota is defined as.

$q_i = q \bullet \sum_j w_{ji}$, where q is a share of the sum of weights.

i.e., the quota is a percentage of the sum of the weights entering the node. Each node has a different quota value, allowing its individual parameters to be taken into account.

A critical set S for the node i is defined as a set of nodes not exceeding size k entering node i such that

$$\sum_{j \in S} w_{ji} \geq q_i.$$

The parameter k is the number of nodes that can simultaneously influence node i. In this work, $k = 3$ (this choice is due to computational complexity limitations).

Thus, the Bundle Index (BI) can be defined as the number of critical sets for node i not exceeding size k.

Bundle index (BI) is a group influence index. For all critical sets S, $|S| \leq k$

$$BI_i(S) = \begin{cases} 1, \textit{if} \sum_{j \in S} w_{ji} \geq q_i \\ 0, \textit{else} \end{cases}$$

BI for a node equals the sum of index values over all its critical sets,

$$BI(i) = \sum_{S} BI_i(S)$$

i.e., $BI(i)$ equals the number of critical sets of node i of size no greater than k.

The second centrality index, PI (Pivotal Index), shows the number of pivotal nodes for each critical set. To introduce this index, it is necessary to define what constitutes a pivotal node. A node j_p is called a pivotal node for node i in a critical set S if $\sum_{j \in S} w_{ji} \geq q_i$ и $\sum_{j \in S \setminus \{j_p\}} w_{ji} < q_i$, i.e. the total incoming weight is no less than the quota (S is critical), but upon removing the node j_p from the set, the sum falls below the quota (S ceases to be critical). For a specific critical set, the value of the $PI_i(S)$ index is equal to the number of pivotal nodes in it. The PI value for node i is the sum over all critical sets of the product of the size of the set $|S|$ and the $PI_i(S)$ value for this set.

$$PI(i) = \sum_{S} |S| \times PI_i(S)$$

The combined index, called the Total influence index (TI) is defined as follows.

$$TI(i) = \alpha_1 IN(i) + \alpha_2 BI(i) + \alpha_3 PI(i.)$$

Here α_i, i =1,2,3 are weighting coefficient, $\alpha_i \in [0,1]$, $\alpha_1 + \alpha_2 + \alpha_3 = 1$. As before the notation for indices is the same, IN-Weighted in-degree, BI- Bundle index, PI-Pivotal index. Throughout the book the equal weights are used, i.e. $\alpha_i = \frac{1}{3}$.

Let us consider the described indices using an example of a graph consisting of five nodes (Fig. 3.3). The adjacency matrix of this graph is shown in Table 3.2, and the values of the In-degree index are shown in Table 3.3.

Let the quota be $q = 1$, $k = 3$. Then for node e we obtain the following critical sets: $\{a\}$, $\{d\}$, and $\{a, d\}$, with total weights 3, 6, and 9 respectively.

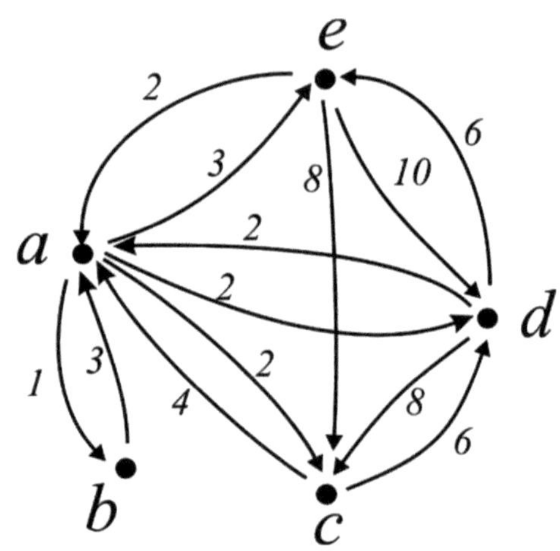

Fig. 3.3 An example of a graph consisting of five vertices

Table 3.2 Adjacency matrix for a graph consisting of five vertices

	a	b	c	d	e
a	0	1	2	2	3
b	3	0	0	0	0
c	4	0	0	6	0
d	2	0	8	0	6
e	2	0	8	10	0

Table 3.3 The values of the In-degree index for a graph consisting of five nodes

Vertices	In-degree
a	0.19
b	0.02
c	0.32
d	0.32
e	0.16

Now the *BI* index for these critical sets can be calculated. Since for all critical sets the sum of weights is greater than or equal to the quota q, *BI* for these sets equals 1, hence $BI(e) = 3$. Similar calculations can be performed for the remaining nodes to obtain the index values (Table 3.4).

To calculate the PI index, it is necessary to identify the pivotal nodes. For node e in the first two critical sets, nodes a и d are pivotal, since their exclusion causes the sum of incoming edge weights to fall below the quota. For the critical set $\{a, d\}$, there are no pivotal nodes, as removing any one node still leaves the sum above the quota. Thus, $PI(e) = 2$. Table 3.4 also presents the index values for the remaining nodes.

Table 3.4 The values of the BI and PI index for a graph consisting of five nodes

Vertices	BI, q = 1, k = 3	PI, q = 1, k = 3
a	0.44	0.31
b	0.03	0.08
c	0.22	0.23
d	0.22	0.23
e	0.09	0.15

The values of the all four indices are presented in Table 3.5.

As can be seen from the description of the indices, a significant difference of the newly proposed centrality indices BI and PI lies in the fact that they take into account the parameters of the vertices, as well as the fact that the centrality of the vertices in the network is determined by the group connections of the given vertex with the others.

Let us consider an example of a network shown in Fig. 3.3. Let us consider the calculation of BI and PI indices for various quota values under the same restriction on the size of critical sets $k = 3$. These new index values are presented in Tables 3.6, 3.7 and 3.8.

Let us compare the values of the BI indices for various quotas with the values of the Weighted In-degree index presented in Table 3.9.

In the Weighted In-degree index, the most influential vertices are *c* and *d*.

For quota $q = 10$, according to the BI index, the most significant vertices are *c* and *d*, and the remaining vertices have the same rank.

When the quota is decreased to 9 ($q = 9$), vertices *c* and *d* have a high rank, but the significance of the other vertices has changed. The ranking now is as follows

Table 3.5 The values of the In-degree, BI, PI, and TI indices for a graph consisting of five nodes

Vertices	In-degree	BI, q = 1, k = 3	PI, q = 1, k = 3	TI
a	0.19	0.44	0.31	0.27
b	0.02	0.03	0.08	0.06
c	0.32	0.22	0.23	0.26
d	0.32	0.22	0.23	0.26
e	0.16	0.09	0.15	0.15

Table 3.6 Values of BI and PI centrality indices for quota $q = 1$

Vertices	BI, q = 1	PI, q = 1
a	0.44	0.31
b	0.03	0.08
c	0.22	0.23
d	0.22	0.23
e	0.09	0.15

Table 3.7 Values of BI and PI centrality indices for quota $q = 9$

Vertices	BI, q = 9	PI, q = 9
a	0.18	0.43
b	0	0
c	0.36	0.29
d	0.36	0.19
e	0.09	0.10

Table 3.8 Values of BI and PI centrality indices for quota $q = 10$

Vertices	BI, q = 10	PI, q = 10
a	0	0
b	0	0
c	0.5	0.6
d	0.5	0.4
e	0	0

Table 3.9 Values of the In-degree centrality index

Vertices	Weighted In-degree
a	0.19
b	0.02
c	0.32
d	0.32
e	0.16

$$c \backsim d \succ a \succ e \succ b,$$

When the quota is reduced to 1 (q =1), the ranking changes completely

$$a \succ c \sim d \succ e \succ b,$$

i.e., the most important vertex becomes a.

In terms of article citations at the journal level, this example can be interpreted as follows. For quota $q = 1$, i.e., if we simply consider the fact of citation of journals, then the most important journal turns out to be a, and the influence of a is twice as high as that of c and d. However when the quota is increasing, journals c and d, actively citing each other, turn out to be the most influential. In other words, the considered models allow us to identify groups of narrowly specialized journals that actively cite each other. Such an effect for economic journals was identified in (Aleskerov et al., 2021).

3.3.2 Network Stability Analysis

In networks considered dynamically, various changes may occur—new vertices may appear or old ones may disappear, and the interconnections between them may change. It is important to understand how much the network has changed over time to identify patterns and trends, and the stability of the scientific community. The simplest methods are the calculation of the correlation of two successive adjacency matrices or the ranks of vertices in such networks (Segarra & Ribeiro, 2016), which do not account for topological changes in the network. In (Aleskerov & Shvydun, 2017), an approach is presented to measure graph stability, taking into account topological similarity and the similarity of the importance of elements. Two networks are considered similar if they have a similar structure (vertices exert the same influence on each other) and similar key elements.

Interval orders

Networks have similar key elements if their vertices have the same ranks. Vertices can be ranked using any centrality indices suitable for the task. Let C_i be the rank of vertex *i*, *n*—the number of vertices, then the interval order matrix $R^t = \left[r_{ij}^t\right]$ is introduced at time *t*

$$r_{ij}^t = \begin{cases} 1, \mathit{if}\ \left|c_i^t - c_j^t\right| > \varepsilon \\ 0, \mathit{else} \end{cases}$$

That is, its values are equal to 1 if the ranks of the vertices differ by more than ε. Then the distance between matrices at two successive points in time can be calculated (Wiener, 1914; Aleskerov et al., 2007)

$$d\left(R^t, R^{t+1}\right) = \frac{\sum_{i \neq j}^{n} \left|r_{ij}^t - r_{ij}^{t+1}\right|}{n \bullet (n-1)}$$

It is equal to zero in the case of matching ranks and greater than zero when the ranks differ. Thus, for example, the ranking of vertices from the above example (Table 3.2) according to the In-degree index, if $\varepsilon = 0$, looks as follows (Table 3.10).

However, if the interval width is increased to $\varepsilon = 0.04$, then vertices *a* and *e* will be assigned the same rank, since their centrality values, 0.16 and 0.19, differ by no more than 0.04. (Table 3.11).

Table 3.10 Ranking of vertices by the in-degree index at $\varepsilon = 0$

Vertex	Rank
c, d	1
a	2
e	3
b	4

Table 3.11 Ranking of vertices by the in-degree index at $\varepsilon = 0.04$

Vertex	Rank
c, d	1
a, e	2
b	3

Topological similarity

Let $\tilde{c}_{ij}^t$ be the influence of vertex i on vertex j at time t. The influence can be described in different ways; in this work, it is proposed to use the share of weights from vertex i to vertex j relative to the total sum of weights of arcs, incoming to vertex j, that is, $\tilde{c}_{ij}^t = \frac{w_{ij}}{\sum_k w_{kj}}$. Then the distance between influence matrices can be calculated as follows

$$\mu\left(\tilde{C}_t, \tilde{C}_{t+1}\right) = \frac{\sum_{i,j}^n \left|\tilde{c}_{ij}^t - \tilde{c}_{ij}^{t+1}\right|}{\mathrm{n}^2 \bullet \gamma},$$

where $\gamma = \max_{i,j}\left(\tilde{c}_{ij}^t, \tilde{c}_{ij}^{t+1}\right)$.

Similarly, for identical graphs, the distance is equal to 0, and for different ones is greater than 0.

For example, let the matrix for the previously considered graph at time t have the form shown in Table 3.2, and at time $t + 1$, the weight on arc (c, a) equals 1. Then in the numerator $\sum_{i,j}^n \left|\tilde{c}_{ij}^t - \tilde{c}_{ij}^{t+1}\right|$ the difference is non-zero only for $j = a$, and the entire sum in the numerator equals 0.48. Thus, $\mu\left(\tilde{C}_t, \tilde{C}_{t+1}\right) = 0.48/25 = 0.02$, since $\gamma = 1$. This reflects the fact that the weight changed for only one arc of the matrix, and the two networks are topologically similar.

Similarity measure

The final similarity metric takes into account how the centralities of the network changed and how its structure changed. Let G^t be graph G at time t, and G^{t+1} at time $t + 1$, and R^t and $\tilde{C}_t$ be the interval order and influence matrices at time t. Then the distance between graphs G^t and G^{t+1}, taking into account both topological similarity and closeness in centralities, is calculated by the formula

$$d\left(G^t, G^{t+1}\right) = \sqrt{\frac{d\left(R^t, R^{t+1}\right)^2 + \delta\left(\tilde{C}_t, \tilde{C}_{t+1}\right)^2}{2}}$$

The closer this value is to 0, the more stable the network over time.

Here $d(R^t, R^{t+1}) = \frac{\sum_{i \neq j=1}^n \left|r_{i,j}^t - r_{i,j}^{t+1}\right|}{n(n-1)}$, $r_{i,j}^t = 1$, if vertex i is above vertex j in the centrality ordering, and $r_{i,j}^t = 0$ otherwise.

Let the modified matrix at time $t + 1$ have the form shown in Table 3.12.

Table 3.12 Modified adjacency matrix for time $t + 1$

	a	b	c	d	e
a	0	1	2	2	3
b	3	0	0	0	0
c	1	0	0	6	0
d	2	0	8	0	6
e	2	0	8	10	0

Table 3.13 Vertex centrality by In-degree index for the modified matrix for time $t + 1$

Vertex	In-degree	Rank
a	0.15	3
b	0.02	4
c	0.33	1
d	0.33	1
e	0.17	2

The centrality values according to the In-degree index are as follows (Table 3.13).

Then the vertex ranks at time $t + 1$ coincide with the ranks at time t, and thus the similarity measure

$$d(G^t, G^{t+1}) = \sqrt{\frac{0.0004}{2}} = 0.014.$$

3.3.3 *Pattern Analysis*

Various clustering methods can be applied to bibliometric data to identify specific groups of objects. We used the pattern analysis method described in (Aleskerov & Alper, 2000; Aleskerov et al., 2013).

Pattern analysis is the analysis of objects grouped into clusters with the same pattern. A pattern is a sample or template; in this work, a pattern is understood as a set of objects characterized by a certain combination of indicator values of the objects. Groups of similar objects will be called clusters. Objects in one cluster are similar to each other and significantly differ from others. Visually, the pattern can be represented by depicting objects (vectors) in parallel coordinates: each vector value is located on its own vertical axis, and all values of one vector are connected by a line for four objects (Fig. 3.4).

To apply the pattern-based clustering algorithm, it is necessary to bring the data to the same scale, i.e., normalize them using any normalization method.

Data normalization is the process of transforming feature values into a unified scale so that each feature contributes equally to the analysis. This is important when using methods that are sensitive to the scale of the data, such as clustering. In one

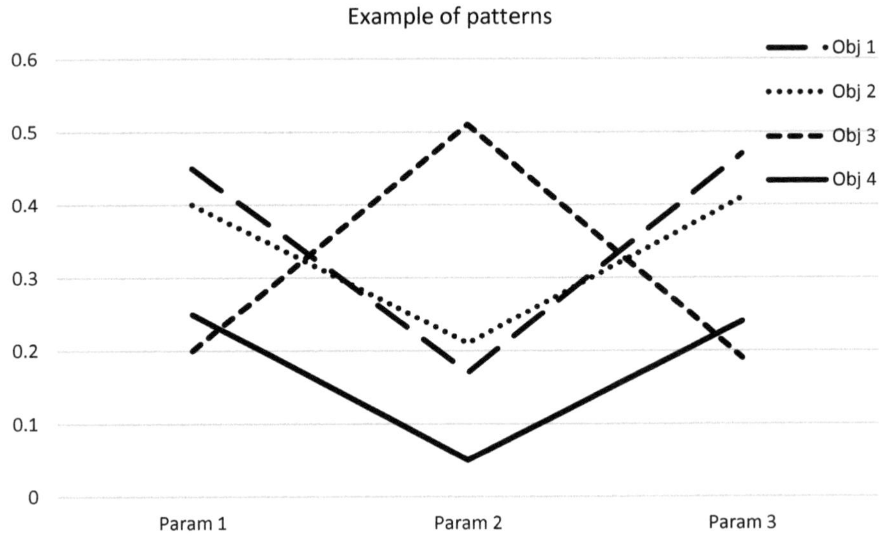

Fig. 3.4 Example of patterns

dataset, different features may have different ranges of values, which can affect the analysis result. To avoid this problem, the data must be brought to the same scale, i.e., normalized.

One of the popular normalization methods is min-max normalization. It involves transforming the values of features into the range from 0 to 1. The following formula is used

$$\frac{x - x_{min}}{x_{max} - x_{min}},$$

where x is the original value of the parameter, x_{min} and x_{max} are the minimum and maximum values of the feature in the dataset. This method preserves the relative differences between the values and makes them comparable with each other.

There are two ways to group into clusters: grouping identical objects with and without considering the scale. In the first case, we get clusters $\{1, 2\}$, $\{3\}$, $\{4\}$, and in the second—$\{1, 2, 4\}$, $\{3\}$ because object 4 has a similar structure but differs in scale. Identical objects are defined by falling into the same neighborhood, called the ε-tube. For the two methods, it is constructed differently:

1. The ε -tube is built on normalized data. Let $v \in \mathbb{R}^k$ be a data vector. In this case, objects that are close in values of vector elements fall into one pattern.

2. The ε -tube is built on the tangents of the angles of polyline slopes. Let $v \in \mathbb{R}^{k-1}$ be a vector of data slope angles. Clustering by tangents combines into one pattern objects whose graphs have a similar shape (points connected by segments in the parallel coordinates system can be represented as a graph of a piecewise linear function).

For each vector v, its ε-tube is calculated, for all v_i the maximum deviation $(d_{max})_i$ is calculated, which is of two types:

- Non-adaptive, does not depend on the value of the parameter $(d_{max})_i = 1$.
- Adaptive, calculated based on the parameter $(d_{max})_i = v_i$.

After that, all values are multiplied by ε: $d = \varepsilon \bullet d_{max}$. Then, if the values $N(v)$ are in the same cluster with vector v, then

$$u \in N(v) \iff \forall i : |u_i - v_i| < d_i .$$

That is, if for each coordinate with number i of vector u the condition is met that the value of this coordinate differs from the value of the corresponding coordinate of vector v by no more than d_i, then vector u falls into the cluster formed by vector v.

Thus, in the considered example shown in Fig. 3.4, for each object we introduce an ε-tube, as shown in Fig. 3.5.

As can be seen, for objects numbered 1 and 2 the ε -tubes intersect, i.e., these objects are placed in one cluster, while objects numbered 3 and 4 each form their own cluster.

Thus, it is possible to group objects into various clusters by choosing the appropriate method for accounting for their structure and scale.

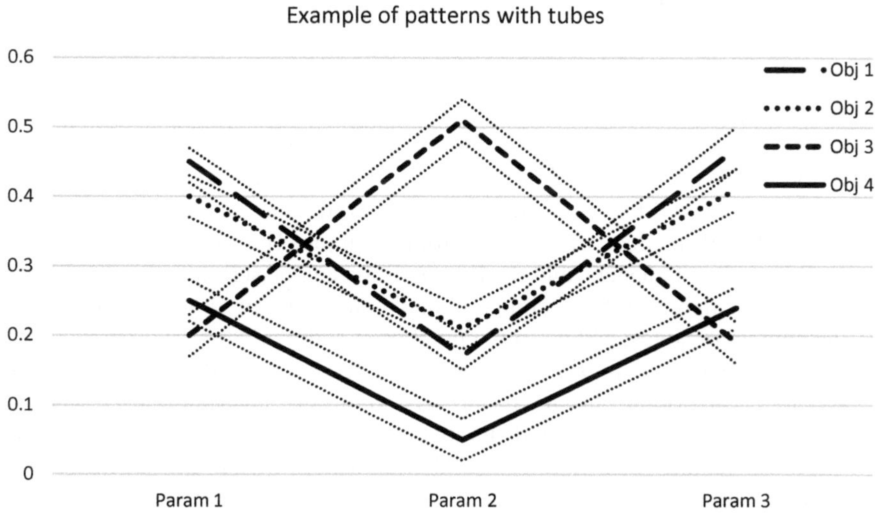

Fig. 3.5 Example of patterns with ε-tube

3.3.4 Community Detection

A useful tool in network analysis is community detection. Community or cluster detection allows grouping objects that are similar in certain characteristics. For citation networks, this approach can be used to identify groups of actively interacting organizations or articles that frequently cite each other and may belong to the same topic. Good communities exhibit high internal density and reachability of nodes, and weak connectivity with the rest of the graph. Accordingly, clustering methods aim to optimize functions corresponding to these criteria.

A classical metric for evaluating graph partitioning into communities is modularity (Newman, 2006). It shows how well nodes are distributed among classes compared to a random partition. The metric can take values from -1 to 1; its maximization allows identifying the best partition. Modularity is described by the following formula

$$Q = \frac{1}{2m} \sum_{i,j \in V} \left[A_{ij} - \frac{k_i k_j}{2m} \right] \delta\left(c_i, c_j\right),$$

where m—the sum of weights of all edges; A_{ij}—the adjacency matrix with edge weights; k_i, k_j—the sum of weights of edges adjacent to nodes i and j; c_i, c_j—the classes of nodes i и j; δ—the Kronecker delta function.

There exist many clustering methods with different specificities and levels of complexity. In network analysis, several methods of community detection were considered

1. *Girvan-Newman* (Newman & Girvan, 2004)—a hierarchical clustering algorithm. At each step of the algorithm, the betweenness centrality of all edges is computed (i.e., the number of shortest paths between nodes that pass through a given edge), and the edge with the highest betweenness is removed. Steps are repeated until all edges are removed from the graph. The output of the algorithm is a hierarchy of clusters, and the best partition can be selected based on modularity. This algorithm has a computational complexity of $O(m^2n)$, where m—is the number of edges, and n—is the number of nodes.
2. *Label Propagation Algorithm* (Cordasco & Gargano, 2010)—an information spreading algorithm. In the basic version of the algorithm, each node is assigned a separate class, and iteratively classes are reassigned by choosing the most popular class among neighboring nodes. When this algorithm was applied to the graph, one large community encompassing almost all nodes and many small communities were identified; the communities resembled connected components.
3. *Louvain method* (Blondel et al., 2008)—a two-phase algorithm based on modularity maximization. In the first phase, each node is assigned a separate community, and modularity is iteratively maximized by moving nodes into other communities. In the second phase, nodes in the same community are grouped and become nodes in a new graph, to which the first phase is reapplied. Due to the two-phase

approach, this algorithm enables identifying the optimal number of communities and maximizing modularity.

4. *Smart Local Moving Algorithm* (Waltman & van Eck, 2013), which underlies the Leiden algorithm (Traag et al., 2019), used in the bibliometric data analysis tool (CitNetExplorer) and for visualization (VosViewer). The algorithm is iterative, and at each step the following actions are performed: optimal communities are identified by modularity as in the first phase of the Louvain algorithm; for each detected community, a similar partitioning is conducted; the resulting community partitions are grouped as in the second phase of the Louvain algorithm. These steps are repeated for the grouped graph until the optimal partition is found. The authors of the method demonstrated in experiments that the algorithm outperforms others in quality and detects communities more accurately. The resulting communities are connected and optimal.

3.4 Algorithm for Network Data Analysis

Thus, the algorithm for analysis using network models consists of the following steps

1. Defining the boundaries of the subject area using keywords corresponding to the field of research interest.
2. Downloading data from the scientific publication database based on keywords in the title, abstract, or keyword section.
3. Data preprocessing: removing irrelevant or erroneous publications, removing publications that do not contain critically important attributes for the analysis of interest (for example, information about the journal). At the preprocessing stage, returns from subsequent stages are possible, because the use of network models makes it possible to identify features in the data already during the process.
4. Constructing a network, for example, a citation network for popularity analysis, an interaction network for community analysis, a term network for semantic similarity analysis, an independent citation network for objective popularity assessment. Networks can be weighted and unweighted, directed and undirected, and take into account different time intervals.
5. Processing the resulting network: removing incorrect edges, isolated nodes, nodes not from the original set. At this step, the analysis of connected components can help, because isolated components often turn out to be data errors. Also, for some tasks, it is possible to eliminate part of the edges or nodes so that the subsequent analysis is indicative. For example, if the graph has many connections, a certain percentage of weak edges can be removed or cores can be extracted in the graph.
6. Calculation of centrality indices, identification of the most significant nodes, and their analysis from the perspective of the subject area.
7. Analysis of network stability if the graph has a temporal structure.

8. Conducting pattern analysis, for example, for changes in centrality indices by year.
9. Identification and analysis of communities in the graph. Such communities are characterized by strong ties within the community, a certain set of nodes, and relatively weak ties between different communities.

Using the example of publications about various aspects of Parkinson's disease, the following chapters will consider the results of applying the described network analysis algorithm.

References

Aleskerov, F., & Alper, C. (2000). A clustering approach to some monetary facts: A long-run analysis of cross-country data. *The Japanese Economic Review, 51*(4), 555–567. https://doi.org/10.1111/1468-5876.00170

Aleskerov, F., Bouyssou, D., & Monjardet, B. (2007). *Utility Maximization, Choice and Preference.* Springer Berlin, Heidelberg, pp. XII, 283. ISBN 978-3-540-34182-6. doi: https://doi.org/10.1007/978-3-540-34183-3.

Aleskerov, F., & Shvydun, S. (2017). A mathematical approach to conflict resolution in the arctic region. In *3rd IEEE international conference on cybernetics (CYBCON)* (pp. 145–150). Exeter.

Aleskerov, F., & Yakuba, V. (2020). Matrix-vector approach to construct generalized centrality indices in networks. *SSRN, 3597948*, ssrn.com/abstract=3597948.

Aleskerov, F. T., Belousova, V. Y., Egorova, L. G., & Mirkin, B. G. (2013). Analysis of patterns in statics and dynamics, part 2: Examples of application to the analysis of socio-economic processes. *Business informatics, 4*(26), 3–20. (in Russian).

Aleskerov, F. T., Kazachinskaya, A. I., Karabekyan, D. S., Semina, A. S., & Yakuba, V. I. (2021). Economic journals of Russia, their characteristics and network analysis. *Journal of the New Economic Association, 50*(2), 170–182. https://doi.org/10.31737/2221-2264-2021-50-2-9(in Russian).

Blondel, V. D., Guillaume, J. L., Lambiotte, R., & Lefebvreet, E. (2008). Fast unfolding of communities in large networks. *Journal of Statistical Mechanics: Theory and Experiment, 10*, P10008. https://doi.org/10.1088/17425468/2008/10/P10008

Bonacich, P. (1972). Factoring and weighting approaches to status scores and clique identification. *Journal of Mathematical Sociology, 2*(1), 113–120.

Brin, S., & Page, L. (1998). The anatomy of a large-scale hypertextual web search engine. Computer networks and ISDN systems, 30 (1–7), 107–117.

Cordasco, G., & Gargano, L. (2010). Community detection via semi-synchronous label propagation algorithms. *IEEE international workshop on: business applications of social network analysis (BASNA)*, IEEE. 1–8. doi: https://doi.org/10.1109/BASNA.2010.5730298.

Freeman, L. C. (1977). A set of measures of centrality based on betweenness. *Sociometry, 40*(1), 35–41.

Newman, M. E. (2006). Modularity and community structure in networks. *Proceedings of the National Academy of Sciences, 103*(23), 8577–8582. https://doi.org/10.1073/pnas.0601602103

Newman, M. E. (2010). *Networks: An introduction.* Oxford University Press.

Newman, M. E., & Girvan, M. (2004). Finding and evaluating community structure in networks. *Physical Review E, 69*(2), 026113. https://doi.org/10.1103/PhysRevE.69.026113

Segarra, S., & Ribeiro, A. (2016). Stability and continuity of centrality measures in weighted graphs. *IEEE Transactions on Signal Processing, 64*(3), 543–555.

Sinha, A., Shen, Z., Song, Y., Ma, H., Eide, D., Hsu, B.-J., & Wang, K. (2015). An overview of Microsoft academic service (MAS) and applications. *Proceedings of the 24th international world wide web conferences,* 243–246. https://doi.org/10.1145/2740908.2742839

Traag, V. A., Waltman, L., & van Eck, N. J. (2019). From Louvain to Leiden: Guaranteeing well-connected communities. *Scientific Reports, 9*(1), 5233. https://doi.org/10.1038/s41598-019-41695-z

Waltman, L., & van Eck, N. J. (2013). A smart local moving algorithm for large-scale modularity-based community detection. *Eur. Phys. J., 86*, 471. https://doi.org/10.1140/epjb/e2013-40829-0

Wang, K., Shen, Z., Huang, C., Wu, C.-H., Eide, D., Dong, Y., Qian, J., Kanakia, A., Chen, A., & Rogahn, R. (2019). A review of Microsoft academic services for science of science studies. *Frontiers in Big Data, 2*, 45. https://doi.org/10.3389/fdata.2019.00045

Wiener, N. (1914). A simplification in the logic of relations. *Proceedings of the Cambridge Philosophical Society, 13*, 387–390.

Chapter 4
An Analysis of Publication Citations

One of the most important assessments of scientific productivity is citation metrics of publications. These indicators determine the relevance of publication topics, their demand, and are quite a significant factor in assessing the scientific productivity and influence of authors, scientific teams, and organizations as a whole.

To increase the accuracy of bibliometric analysis, new methods of analysis and modification of indicators are being developed. Thus, in (Aroeira, R 2020), the accuracy of bibliometric indicators such as the Hirsch index, journal impact factor, and total number of citations in assessing the real scientific contribution of publications is analyzed. The authors propose a new method of citation analysis that makes it possible to identify those references that actually use ideas, methods, or hypotheses to create new knowledge. The methodology is based on a sample of publications and a detailed analysis of their citations, which makes it possible to distinguish formal references from those that indicate real scientific impact. The data obtained show that only about 12.4% of citations reflect the real contribution of an article to the development of science, and a high journal impact factor does not guarantee the significance of a work.

In (Pagani R & Resende 2015), the Methodi Ordinatio is proposed, which allows for the selection and ranking of scientific publications, taking into account the journal impact factor, number of citations, and year of publication. The method is based on the integration of bibliometric indicators into a single index that makes it possible to identify the most significant publications in a particular field. Unlike the traditional approach, which may overestimate old works with a large number of citations or underestimate recent studies, Methodi Ordinatio takes the time factor into account.

The study (Su & Chang 2024) deals with the analysis of the citation network of scientific publications on the problem of root caries in the elderly. The main focus is on identifying the most significant publications, key research directions, and gaps in the existing literature. The co-citation method makes it possible to determine which articles are most frequently cited, indicating their importance in shaping the scientific discussion in this field. During the analysis, a group of key publications

© The Editor(s) (if applicable) and The Author(s), under exclusive license to Springer Nature Switzerland AG 2026

F. Aleskerov et al., *Bibliometric Analysis by Network Models*, Contemporary Systems Thinking, https://doi.org/10.1007/978-3-032-09171-0_4

focused on the fields of dentistry, maxillofacial surgery, and medicine was identified, emphasizing the interdisciplinary nature of research on this topic.

In (Fan, 2024), a bibliometric and network analysis of scientific publications on COVID-19 for 2020 is conducted, exploring trends, author collaborations, and key topics. Metrics such as h-index, g-index, as well as classical centrality indices, are used for the analysis. Using network analysis methods, co-authorship, institutional, and keyword co-occurrence networks are examined.

This chapter presents a network analysis of the citation graph of publications using classical and new centrality indices and provides an analysis of the network of independent citations of articles.

4.1 Citation Graph of Publications

The citation graph of publications makes it possible to analyze the interaction and influence exerted by studies presented in scientific publications on each other. The nodes in this graph are articles, and the directed unweighted edges are citations (all weights are equal to 1). Thus, if an article with ID1 cites an article with ID2 in year Y, then a directed edge from node ID1 to node ID2 appears in the network (Table 4.1). The 'weight' column contains the weight of the edge or the number of citations, i.e., in this network it is equal to 1, because each article can cite another no more than once.

The constructed graph consisted of 94 connected components. A connected component is a part of a graph (a subgraph) representing a group of nodes, each of which is connected to every other node in that group. The largest connected component contained 39,618 articles; the others contained 2–3 articles citing each other. For example, in the component 71 there are two publications on the effect of levodopa—a common drug for the treatment of Parkinson's disease. In the component 94–2 articles by the same author. And in the component 17 there are three articles on artifacts in electrocardiography. Such articles are not excluded from the final list, but when calculating centrality indices, they will have values close to 0, because they are not significant in the overall network due to the small number of links.

Table 4.1 A fragment of the citation network of articles

ID 1	ID 2	weight	year
2754967293	2339791932	1	2017
2112455323	1920030402	1	2015
2177834950	2123627348	1	2015
2558041282	1934236512	1	2016
2558041282	1947901277	1	2016

Table 4.2 Number of citations per article

Number of articles	Ranges of number of citations
9626	0
22,035	1–10
7302	11–49
642	50–99
166	100–199
39	200–500
1	1563

4.2 Statistical Analysis of Publication Citations

An analysis of the number of citations of the articles included in the constructed network showed that one article had 1563 citations, while 9626 articles had 0 citations. Table 4.2 presents the distribution of the number of citations of articles over the entire period under consideration.

The article that received the highest number of citations in our publication citation network was published in 2015 in the journal "Movement Disorders" (ISSN: 08853185, 15,318,257) under the title "MDS clinical diagnostic criteria for Parkinson's disease" (DOI 10.1002/MDS.26424). The article presents clinical diagnostic criteria for Parkinson's disease (PD) recommended by the MDS (International Parkinson and Movement Disorder Society). The PD criteria are intended for use in clinical research but can also be used for clinical diagnosis. The large number of citations is due to the fact that the article is considered a guideline for diagnosis.

4.3 Classical Centrality Indices

Below we evaluated classical centrality indices for article citations. The value of the In-degree index behaves the same as the weight of the edge, because it equals the number of incoming edges, i.e., the number of citations. Table 4.3 presents the Top-10 publications with the highest number of citations for 2015–2021. The table also lists the journals in which the articles were published and the year of their publication.

For comparison, Table 4.4 presents these Top-10 articles sorted in descending order of the In-degree index, as well as their rank for other classical indices. Lower rank numbers correspond to higher positions in the ranking.

The In-degree index highlighted works with a large number of citations. This list includes articles containing MDS recommendations on diagnostic criteria for various forms of PD, including early diagnosis (No. 1 and 4), and work No. 6 are on the theme of an early diagnosis based on non-motor symptoms. These articles are fundamental for the diagnosis of PD. The list also includes publications on the study of risk factors (No. 3 and 5) and statistical analysis of medical data on PD

Table 4.3 Top-10 publications with the highest citation count

DOI	Title of the article	Year of publ.	Journal	Number of citations
10.1002/MDS.26424	MDS clinical diagnostic criteria for Parkinson's disease	2015	Movement disorders ISSN: 08853185, 15,318,257	1563
10.1016/ J.CELL.2016.11.018	Gut microbiota regulate motor deficits and Neuroinflammation in a model of Parkinson's disease	2016	Cell ISSN: 00928674, 10,974,172	500
10.1007/ S00702-017-1686-Y	Epidemiology of Parkinson's disease.	2017	Journal of neural transmission ISSN: 03009564, 14,351,463	485
10.1002/MDS.26431	MDS research criteria for prodromal Parkinson's disease	2015	Movement disorders ISSN: 08853185, 15,318,257	473
10.1016/ S1474-4422(16)30230-7	The epidemiology of Parkinson's disease: Risk factors and prevention	2016	Lancet neurology ISSN: 14744465, 14,744,422	469
10.1038/NRN.2017.62	Non-motor features of Parkinson disease.	2017	Nature reviews neuroscience, ISSN: 14710048, 1,471,003X	459
10.1016/S1474-4422 (18) 30295-3	Global, regional, and national burden of Parkinson's disease, 1990–2016: a systematic analysis for the Global Burden of Disease Study 2016	2018	Lancet neurology ISSN:14744465, 14,744,422	459
10.1016/ J.NEURON.2014.12.007	The roles of PINK1, parkin and mitochondrial Fidelity in Parkinson's disease	2015	Neuron ISSN:10974199, 08966273	458
10.1038/NATURE14547	α-Synuclein strains cause distinct synucleinopathies after local and systemic administration	2015	Nature ISSN:14764687, 00280836	439
10.1002/MDS.26069	Gut microbiota are related to Parkinson's disease and clinical phenotype.	2015	Movement Disorders ISSN 08853185, 15,318,257	416

Table 4.4 In-degree index values of the Top-10 publications and their rank for other classical indices

In-degree rank	Title	In-degree value	Rank		
			Betweenness	Eigenvector	PageRank
1	MD Sclinical diagnostic criteria for Parkinson's disease	0.005	1907	15	1
2	Gut microbiota regulate motor deficits and Neuroinflammation in a model of Parkinson's disease	0.0016	118	62	8
3	Epidemiology of Parkinson's disease	0.00156	73	1224	5
4	MDS research criteria for prodromal Parkinson's disease	0.00152	799	13	3
5	The epidemiology of Parkinson's disease: Risk factors and prevention	0.0015	57	705	4
6	Non-motor features of Parkinson disease	0.00148	69	764	14
7	Global, regional, and national burden of Parkinson's disease, 1990–2016: a systematic analysis for the Global Burden of Disease Study 2016	0.00148	245	3585	7
8	The roles of PINK1, parkin and mitochondrial Fidelity in Parkinson's disease	0.00147	21,387	4	2
9	α-Synuclein strains cause distinct synucleinopathies after local and systemic administration	0.0014	2735	1	6
10	Gut microbiota are related to Parkinson's disease and clinical phenotype	0.0013	21,387	2	10

(No. 7). The In-degree index also highlighted three publications (No. 2, 8, 10). They are about biochemical and genetic research, which may indicate that these research areas are in demand and attract great interest in the scientific community.

As can be seen from Table 4.4, the leaders in citation count (In-degree index) do not occupy leading positions in the ranked list by Betweenness and Eigenvector indices, because they do not reflect the degree of article connectivity. PageRank, on the contrary, shows similar results.

In terms of the Betweenness index, articles that lie on the shortest paths between other articles are of great importance. They may not have a large number of citations or important neighbors, but still play the role of intermediaries. Table 4.5 presents the Top-10 articles by Betweenness index value, as well as the rank of these publications by other classical indices.

In Table 4.5, a number of new articles appear in the Top-10 that were not present in the previous table (see Table 4.4). The Top-10 publications by the Betweenness index are dominated by biochemical, genetic, and clinical studies. For instance, articles (No. 1, 3, 4, 9) highlight the theme of the biochemical aspects of disease causes; publications No. 5 and 10—to the differential diagnosis of diseases with similar symptoms. Articles No. 2 and 6 consider various approaches to disease treatment. Thus, the Betweenness index made it possible to highlight articles that are important connecting separate subgroups of publications.

The Eigenvector index allows assessing the importance (popularity) of a publication, which is proportional to the importance of the publications it is connected to. Table 4.6 presents the Top-10 articles by Eigenvector index value, as well as the rank of these publications by other classical indices. The majority of the studies presented in Table 3.6 relate to biochemical and genetic research (No. 1, 2, 4, 6, 8, 9, 10); the article No. 7 deals with construction of a model for biochemical research. The publications No. 3 and 5 are about clinical studies aimed at PD treatment. Articles No. 1, 2, and 4 are also in the Top-10 by the In-degree index. It should also be noted that the Betweenness index values for all articles from this list are among the lowest.

Table 4.7 contains articles that had high importance in terms of the PageRank index, in which the index value of an article gives importance to its neighbors proportionally to their number. That is, if an important article cites dozens of other articles, this does not mean that their importance should also increase greatly. The list of publications is similar to the list of the most cited publications with the highest In-degree value (Table 4.4). New article 9 appears on the field of neuroscience.

We analyzed the dynamics of changes in centrality indices for the considered time period 2015–2021. Table 4.8 presents the ranks for the In-degree index in each considered year.

Depending on the month of publication, an article may not have any references in the year of publication or have a small number of references, which is reflected in its rank. In the table, the absence of references is marked with "-", high rank values indicate that the publication had a small number of references in that year. Checking the dependence of the number of references to publications on the number of publications per year did not reveal a direct dependence. Figure 4.1 shows a diagram of the number of publications in each year under consideration.

Table 4.5 Betweenness index values of the Top-10 publications and their rank for other classical indices

№	Title	Journal / year of publ.	Betweenness value	Rank		
				In-degree	Eigenvector	PageRank
1	Selective neuronal vulnerability in Parkinson disease	Nat Rev. Neurosci 2017	0.0009	14	131	31
2	Probiotics for Parkinson's disease	International Journal of Molecular Sciences 2019	0.0008	186	6191	6368
3	Can the gut be the missing piece in uncovering PD pathogenesis	Parkinsonism & Related Disorders 2018	0.00075	193	4922	12,172
4	α-Synuclein in Parkinson's disease: Causal or bystander?	Journal of Neural Transmission 2019	0.00074	187	6731	9689
5	Dementia with Lewy bodies and Parkinson's disease-dementia: Current concepts and controversies	Journal of Neural Transmission 2018	0.0007	141	3106	882
6	Disease modification in Parkinson's disease: Current approaches, challenges, and future considerations	Movement Disorders 2018	0.00069	109	2502	555
7	Neuropathology and pathogenesis of extrapyramidal movement disorders: a critical update, hypokinetic-rigid movement disorders	Journal of Neural Transmission 2019	0.00069	202	9042	11,700

(continued)

Table 4.5 (continued)

№	Title	Journal / year of publ.	Betweenness value	Rank		
				In-degree	Eigenvector	PageRank
8	Past, present, and future of Parkinson's disease: A special essay on the 200th anniversary of the shaking palsy future of Parkinson's disease: A special essay on the 200th anniversary of the shaking palsy	Movement Disorders 2017	0.00068	28	1322	83
9	Heterogeneity in α-synuclein subtypes and their expression in cortical brain tissue lysates from Lewy body diseases and Alzheimer's disease	Neuropathology and Applied Neurobiology 2019	0.00064	201	8151	13,906
10	Are dementia with Lewy bodies and Parkinson's disease dementia the same disease?	BMC Medicine 2018	0.00053	147	2711	1041

Thus, the article "MDS clinical diagnostic criteria for Parkinson's disease", describing the clinical diagnostic criteria for Parkinson's disease recommended by the International Parkinson and Movement Disorder Society (MDS), consistently ranks first in terms of the number of citations. Basically, the peak values of citations to articles are reached within 1–3 years after publication, then new articles appear, and the number of citations decreases. Let us give, as an example, the dynamics of changes in the In-degree index values by year for the Top-5 publications listed in Table 4.8 (Fig. 4.2).

The index of article 1 grows for two years after publication and then maintains high values. The indices of other articles also grow for the first 2–3 years after publication, and then slow down or start to fall. This trend is well illustrated by the change in the Eigenvector index over the years. The Eigenvector index evaluates the relative weight of an article, i.e. the degree of importance, which is proportional to the degree of importance of the publications with which it is related. Using the example of the Top-3 articles (1—"α-Synuclein strains cause distinct synucleinopathies after local and

Table 4.6 Eigenvector index values of the Top-10 publications and their rank for other classical indices

№	Title	Journal / year of publ.	Eigenvector value	Rank		
				In-degree	Betweenness	PageRank
1	α-Synuclein strains cause distinct synucleinopathies after local and systemic administration	Nature 2015	0.3445	8	2735	6
2	Gut microbiota are related to Parkinson's disease and clinical phenotype	Movement Disorders 2015	0.227	9	21,387	10
3	Vagotomy and subsequent risk of Parkinson's disease	Annals of Neurology 2015	0.1828	16	21,387	47
4	The roles of PINK1, parkin and mitochondrial Fidelity in Parkinson's disease	Neuron 2015	0.159	7	21,387	2
5	Parkinson risk in idiopathic REM sleep behavior disorder: Preparing for neuroprotective trials	Neurology 2015	0.1579	48	21,387	44
6	Colonic mucosal α-synuclein lacks specificity as a biomarker for Parkinson's disease	Neurology 2015	0.1525	112	21,387	210

(continued)

Table 4.6 (continued)

№	Title	Journal / year of publ.	Eigenvector value	Rank		
				In-degree	Betweenness	PageRank
7	Longitudinal follow-up and characterization of a robust rat model for Parkinson's disease based on overexpression of α-synuclein with adeno-associated viral vectors	Neurobiology of Aging 2015	0.1364	164	21,387	212
8	Evidence for α-synuclein prions causing multiple system atrophy in humans with parkinsonism	Proceedings of the National Academy of Sciences 2015	0.1337	35	4010	28
9	In vivo gastric detection of α-synuclein inclusions in Parkinson's disease	Movement Disorders 2015	0.1336	140	21,387	267
10	Colonic bacterial composition in Parkinson's disease	Movement Disorders 2015	0.1259	15	4270	42

systemic administration"; 2—"Gut microbiota are related to Parkinson's disease and clinical phenotype"; 3—"Vagotomy and subsequent risk of Parkinson's disease". All articles were published in 2015), we will show how the index value changes over the years. All Top-3 articles had the maximum index value in 2017–2018 and dropped sharply in 2020. (Fig. 4.3). This means that their importance has decreased, and other publications have emerged that tend to increase in the Eigenvector index value.

The change in PageRank index values by years is the same as that of the In-degree index. As an illustration, Fig. 4.4 shows the PageRank index values by years for the Top-3 articles (1—"MDS clinical diagnostic criteria for Parkinson's disease"; 2—"The Roles of PINK1, Parkin and Mitochondrial Fidelity in Parkinson's Disease"; 3—"MDS research criteria for prodromal Parkinson's disease"). All articles were published in 2015.

As can be seen from Fig. 4.4, the article "MDS clinical diagnostic criteria for Parkinson's disease" maintains its rank, while the other two articles, having reached the maximum index value in 2016–2017, lose their positions in subsequent years.

Table 4.7 PageRank index value Top-10 publications and their rank for other classical indices

№	Title	Journal /year of publ.	PageRank value	Rank		
				In-degree	Betweenness	Eigenvector
1	MDS clinical diagnostic criteria for Parkinson's disease	Movement Disorders 2015	0.0047	1	1907	15
2	The roles of PINK1, parkin and mitochondrial Fidelity in Parkinson's disease	Neuron 2015	0.0015	7	21,387	4
3	MDS research criteria for prodromal Parkinson's disease	Movement Disorders 2015	0.0011	4	799	13
4	The epidemiology of Parkinson's disease: Risk factors and prevention	Lancet Neurology 2016	0.00109	5	57	705
5	Epidemiology of Parkinson's disease	Journal of Neural Transmission 2017	0.00105	3	73	1224
6	α-Synuclein strains cause distinct synucleinopathies after local and systemic administration	Nature 2015	0.00102	8	2735	1
7	Global, regional, and national burden of Parkinson's disease, 1990–2016: a systematic analysis for the global burden of disease study 016	Lancet Neurology 2018	0.00092	6	245	3585
8	Gut microbiota regulate motor deficits and Neuroinflammation in a model of Parkinson's disease	Cell 2016	0.0009	2	118	62

(continued)

Table 4.7 (continued)

№	Title	Journal /year of publ.	PageRank value	Rank		
				In-degree	Betweenness	Eigenvector
9	Oxidative stress and Parkinson's disease	Frontiers in Neuroanatomy 2015	0.00087	18	755	397
10	Gut microbiota are related to Parkinson's disease and clinical phenotype	Movement Disorders 2015	0.00086	9	21,387	2

Next, we will consider new centrality indices. These are the BI (Bundle Index)—group influence index and PI (Pivotal Index)—influence index by pivotal vertices.

4.4 New Centrality Indices

New centrality indices were applied to identify groups of specialized articles that actively cite each other. The maximum size of the critical set is 3, that is, 3 nodes can simultaneously influence another node. The following indices were calculated: PI, BI, In-degree, and the overall influence index TI. For the new indices, the following quota values were considered as percentages of the total number of citations per article,

- 0.1% of citations
- 0.5% of citations
- 1% of citations.

Higher quota values are not presented because the edge weight is equal to 1, and the sum of incoming edge weights from the critical set cannot exceed the quota.

Table 4.9 presents the values of the PI, BI, TI indices for the 0.1% quota and the In-degree index for the Top-10 articles. The list of publications is ranked by TI.

The list of articles completely matches the Top-10 according to the In-degree index (see Table 4.4). The leading position is held by the first article, which is the most cited among all articles. The values of the PI index (0.94) and the BI index (0.63) for this article are significantly higher than for the other publications.

Figure 4.5 a, b shows the dynamics of changes in the BI and PI indices by year for the Top-5 publications presented in Table 4.9. The Bundle index (Fig. 4.5a) for the first article was the highest after 2016, whereas for the other publications the index was significantly lower and fluctuated between 0 and 0.15, with a maximum value 1–3 years after publication. The dynamics of the Pivotal index by year (Fig. 4.5b) fully coincides with the In-degree index. In this case, the small quota ($q = 0.1\%$) does not lead to a change in the number of pivotal nodes. The value of TI over the

Table 4.8 In-degree ranks by years

In-degree rank (7 years)	Title	Year of publ.	2015	2016	2017	2018	2019	2020	2021
1	MDS clinical diagnostic criteria for Parkinson's disease	2015	–	6	1	1	1	1	1
2	Gut microbiota regulate motor deficits and Neuroinflammation in a model of Parkinson's disease	2016	–	–	8	2	3	7	8
3	Epidemiology of Parkinson's disease	2017	–	–	2247	18	5	3	4
4	MDS research criteria for prodromal Parkinson's disease	2015	188	7	2	7	9	10	18
5	The epidemiology of Parkinson's disease: Risk factors and prevention	2016	–	44	8	10	4	7	9
6	Non-motor features of Parkinson disease	2017	–	–	337	10	4	8	6
7	Global, regional, and national burden of Parkinson's disease, 1990–2016: a systematic analysis for the Global Burden of Disease Study 2016	2018	–	–	–	–	52	2	2
8	The roles of PINK1, parkin and mitochondrial Fidelity in Parkinson's disease	2015	2	2	3	6	12	13	14

(continued)

Table 4.8 (continued)

In-degree rank (7 years)	Title	Year of publ.	2015	2016	2017	2018	2019	2020	2021
9	α-Synuclein strains cause distinct synucleinopathies after local and systemic administration	2015	10	1	4	3	6	21	28
10	Gut microbiota are related to Parkinson's disease and clinical phenotype	2015	3	19	11	8	11	11	12

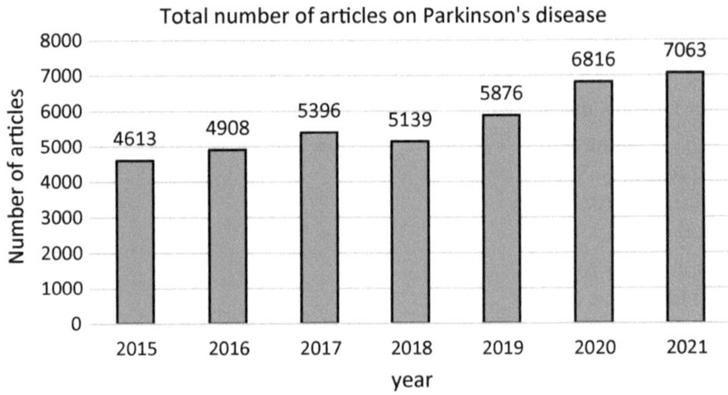

Fig. 4.1 Number of publications by years

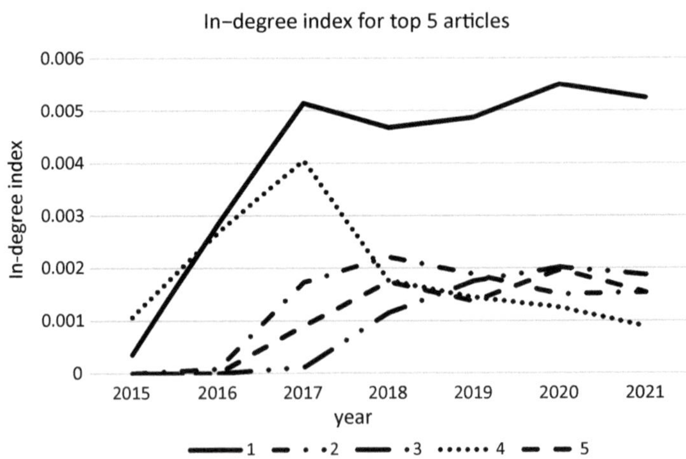

Fig. 4.2 Dynamics of the In-degree index by years for the Top-5 articles

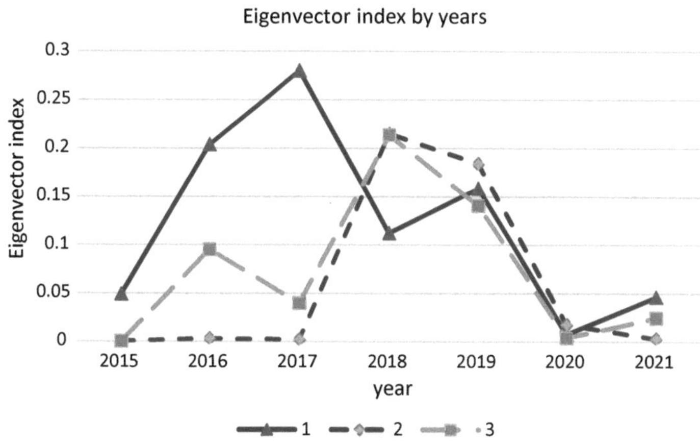

Fig. 4.3 Top-3 articles by Eigenvector index by years

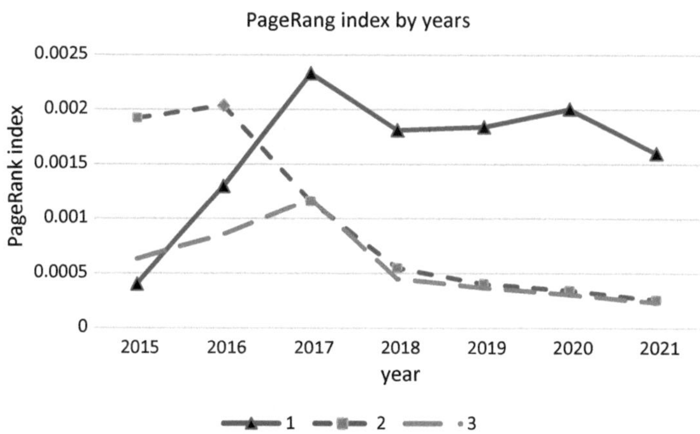

Fig. 4.4 Top-3 articles by PageRank index by years

years behaves similarly to the BI index. Since the numerical value of this index is significantly greater than those of the PI and In-degree indices, its contribution to the overall TI index significantly exceeds the contribution of the PI index.

For the 0.5% quota, the results differ insignificantly: the TI index value for articles with fewer references is already higher (Table 4.10). The most cited articles are lower in the ranking list because increasing the quota reduces the number of critical sets with a sum of weights exceeding the threshold.

A new article appears in the table (No. 10) on genetic research. The changes in BI and PI by year (Fig. 4.6 a, b) show that both indices grow during the first 1–3 years after publication, and then, reaching a peak, their values decrease and may fluctuate slightly.

Table 4.9 Values of PI, BI, TI indices for the 0.1% quota and the In-degree index

№	Title	In-degree	q = 0.1%		
			BI	PI	TI
1	MDS clinical diagnostic criteria for Parkinson's disease	0.005	0.6307	0.9404	0.5254
2	Gut microbiota regulate motor deficits and Neuroinflammation in a model of Parkinson's disease	0.0016	0.0206	0.000096	0.0075
3	Epidemiology of Parkinson's disease	0.00156	0.0188	0.000093	0.0068
4	MDS research criteria for prodromal Parkinson's disease	0.00152	0.0175	0.000091	0.0064
5	The epidemiology of Parkinson's disease: Risk factors and prevention	0.0015	0.01704	0.00009	0.0062
6	Non-motor features of Parkinson disease	0.00147	0.016	0.0000884	0.00584
7	Global, regional, and national burden of Parkinson's disease, 1990–2016: a systematic analysis for the Global Burden of Disease Study 2016	0.00147	0.016	0.0000884	0.00584
8	The roles of PINK1, parkin and mitochondrial Fidelity in Parkinson's disease	0.00147	0.0159	0.0000882	0.00581
9	α-Synuclein strains cause distinct synucleinopathies after local and systemic administration	0.0014	0.014	0.000085	0.0052
10	Gut microbiota are related to Parkinson's disease and clinical phenotype	0.0013	0.0119	0.00008	0.0044

The results for the 1% quota differ significantly (Table 3.11) the most cited articles are no longer present; the list of articles is completely updated. The In-degree index values for these articles are close to zero. The current list includes articles on early diagnostics and clinical studies (No. 1, 2, 4, 6, 9, 10). The remaining articles deal with the biochemical aspects of the disease. Interestingly, with a high quota value, articles with a small number of citations stand out, and they mainly concern clinical studies.

The analysis of BI and PI index values (with a 1% quota) by year showed (Fig. 3.7 a, b) that for the majority of publications, the index values increase in the first 1–2 years after publication, then decrease to a certain level and fluctuate slightly around it. However, there are publications whose index gradually increases, for example, the article "The clinical symptoms of Parkinson's disease," published in 2016 (rank 2 in Table 3.11), showed an increase in the BI index by 2021 (Fig. 3.7a).

An opposite example—a decrease in the index—is shown in Fig. 4.8.

The article "The Onset of Nonmotor Symptoms in Parkinson's disease (The ONSET PD Study)" was published in February 2015 in the journal *Movement Disorders*, and in the same year, the BI and PI index values for No. 6 (Table 4.11) were the highest, while in the subsequent years, the index values decreased.

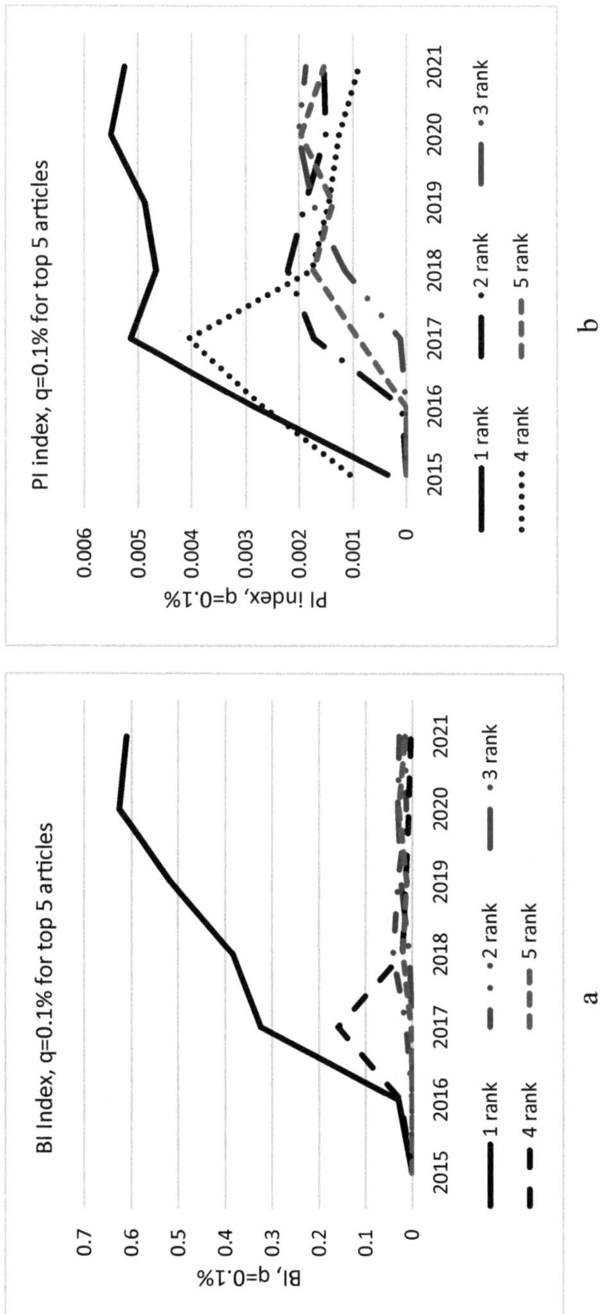

Fig. 4.5 Dynamics of BI (**a**) and PI (**b**) indices ($q = 0.1\%$) by year for the Top-5 publications

Table 4.10 Values of PI, BI, TI indices for the 0.5% quota and the In-degree index

№	Title	In-degree	q = 0.5%		
			BI	PI	TI
1	Gut microbiota regulate motor deficits and Neuroinflammation in a model of Parkinson's disease	0.0016	0.0557	0.1293	0.0622
2	Epidemiology of Parkinson's disease	0.00156	0.0509	0.118	0.0568
3	MDS research criteria for prodromal Parkinson's disease	0.00152	0.0472	0.1094	0.0527
4	The epidemiology of Parkinson's disease: Risk factors and prevention	0.0015	0.046	0.1067	0.0514
5	Non-motor features of Parkinson disease	0.00147	0.043	0.09997	0.0482
6	Global, regional, and national burden of Parkinson's disease, 1990–2016: a systematic analysis for the Global Burden of Disease Study 2016	0.00147	0.043	0.09997	0.0482
7	The roles of PINK1, parkin and mitochondrial Fidelity in Parkinson's disease	0.00147	0.0428	0.0993	0.0479
8	α-Synuclein strains cause distinct synucleinopathies after local and systemic administration	0.0014	0.0377	0.0874	0.0422
9	Gut microbiota are related to Parkinson's disease and clinical phenotype	0.00133	0.032	0.0744	0.0359
10	A met-analysis of genome-wide association studies identifies 17 new Parkinson's disease risk loci	0.00132	0.0314	0.0728	0.0352

The analysis of publications by year revealed several articles with high index values in 2021 that were not among the top in previous years. It can be assumed that the interest of researchers in these articles increased over the past year. Table 4.12 presents a list of these publications. The table shows their ranks by the In-degree index, calculated for the entire period under consideration (2015–2021) without annual division, as well as the name of the index with a high value in 2021 and its rank.

For the In-degree index, three articles appeared that provide general overviews. For the Eigenvector metric, ten articles were identified concerning dopamine, stem cells, and their transplantation. In 2020, one phase of the Transeuro Transplant program, funded by the European Union, concluded, and researchers published their findings. This may open up significant prospects for Parkinson's disease research in this area and reflects the growing interest of the scientific community.

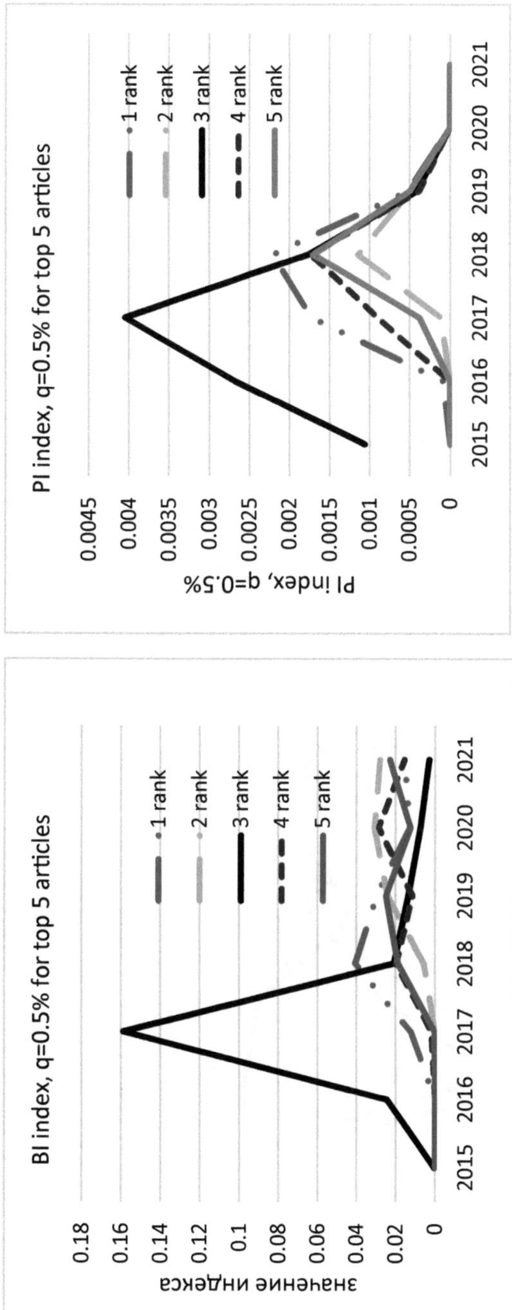

Fig. 4.6 Dynamics of BI (**a**) and PI (**b**) indices (quota q = 0.5%) by year for the Top-5 publications

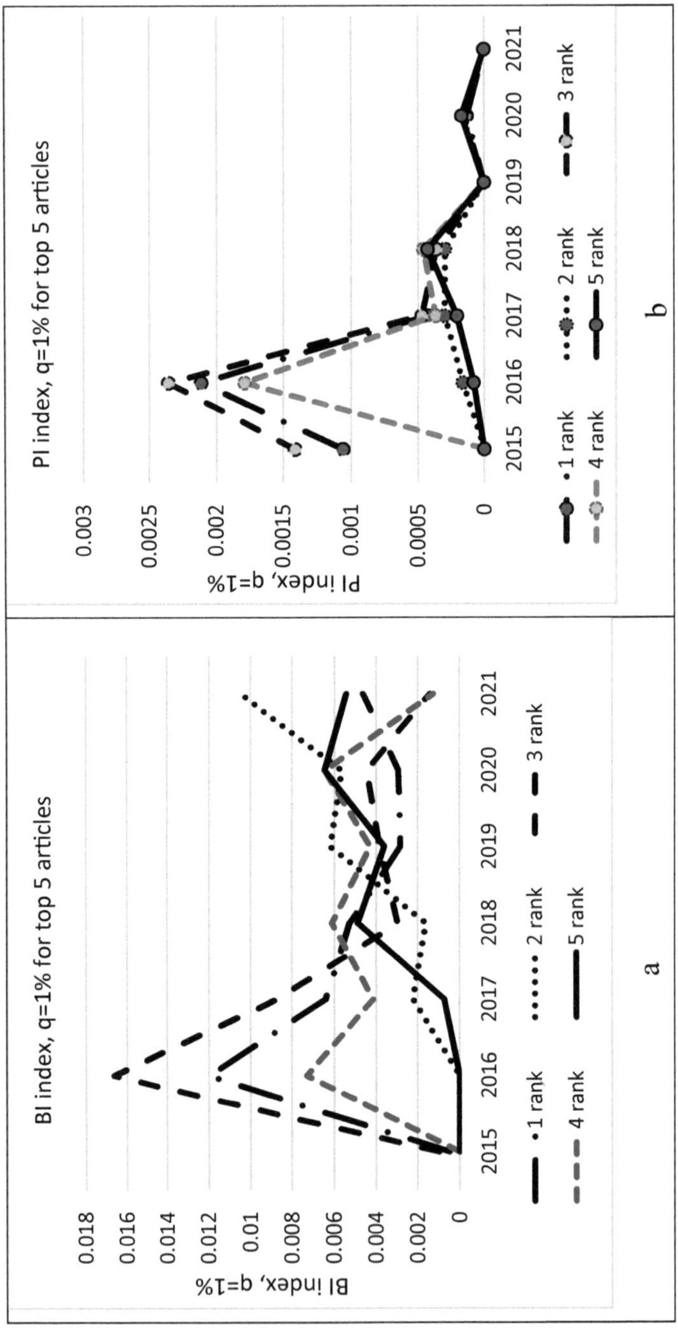

Fig. 4.7 Dynamics of BI (**a**) and PI (**b**) indices (quota q = 1%) by year for the Top-5 publications

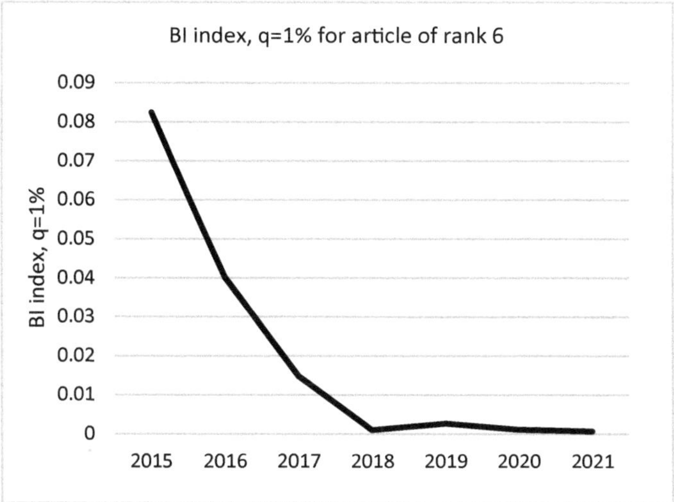

Fig. 4.8 Dynamics of BI index (quota q = 1%) for the article ranked No. 6 (Table 4.11)

Table 4.11 Values of PI, BI, TI indices for the 1% quota and the In-degree index

№	Title	In-degree	q = 1%		
			BI	PI	TI
1	Vagotomy and subsequent risk of Parkinson's disease	0.0009	0.0204	0.068	0.0298
2	The clinical symptoms of Parkinson's disease	0.00089	0.0202	0.0673	0.0295
3	Oxidative stress and Parkinson's disease	0.00087	0.0183	0.061	0.0267
4	Neuroinflammation in Parkinson's disease and its potential as therapeutic target	0.00086	0.0177	0.0589	0.0258
5	Short chain fatty acids and gut microbiota differ between patients with Parkinson's disease and age-matched controls	0.00082	0.0156	0.052	0.0228
6	The Onset of nonmotor symptoms in Parkinson's disease (the ONSET PD study)	0.000814	0.015	0.0501	0.022
7	Transneuronal propagation of pathologic α-Synuclein from the gut to the brain models Parkinson's disease	0.000811	0.0149	0.0496	0.0217
8	T cells from patients with Parkinson's disease recognize α- synuclein peptides	0.00079	0.0135	0.04497	0.0197
9	Prediagnostic presentations of Parkinson's disease in primary care: a case-control study	0.00078	0.0132	0.0439	0.0193
10	New clinical subtypes of Parkinson disease and their longitudinal progression: A prospective cohort comparison with other phenotypes	0.00077	0.012676	0.0422	0.0186

Table 4.12 Top list of the articles of 2021 year, not included in the previous top list

№	Title of the article	Year of publication	In-degree rank in 2015–2021	Index and rank in 2021
1	Identification of novel risk loci, causal insights, and heritable risk for Parkinson's disease: a meta-analysis of genome-wide association studies	2019	12	In-degree 3
2	Diagnosis and treatment of Parkinson disease: A review	2020	32	In-degree 5
3	Global, regional, and national burden of neurological disorders, 1990–2016: A systematic analysis for the Global Burden of Disease Study 2016	2017	31	In-degree 10
4	Extensive graft-derived dopaminergic innervation is maintained 24 years after transplantation in the degenerating parkinsonian brain	2016	108	Eigenvector 1
5	Personalized iPSC-derived dopamine progenitor cells for Parkinson's disease	2020	164	Eigenvector 2
6	Pre-clinical study of induced pluripotent stem cell-derived dopaminergic progenitor cells for Parkinson's disease	2020	182	Eigenvector 3
7	The future of stem cell therapies for Parkinson disease	2020	174	Eigenvector 4
8	Single cell transcriptomics identifies stem cell-derived graft composition in a model of Parkinson's disease	2020	195	Eigenvector 5

(continued)

Table 4.12 (continued)

№	Title of the article	Year of publication	In-degree rank in 2015–2021	Index and rank in 2021
9	Predictive markers guide differentiation to improve graft outcome in clinical translation of hESC-based therapy for Parkinson's disease	2017	132	Eigenvector 6
10	Human trials of stem cell-derived dopamine neurons for Parkinson's disease: Dawn of a new era	2017	111	Eigenvector 7
11	Human iPS cell-derived dopaminergic neurons function in a primate Parkinson's disease model	2017	77	Eigenvector 8
12	Designing stem-cell-based dopamine cell replacement trials for Parkinson's disease	2019	164	Eigenvector 9
13	Single-cell analysis reveals a close relationship between differentiating dopamine and subthalamic nucleus neuronal lineages	2017	185	Eigenvector 10
14	Cognitive decline in Parkinson disease	2017	13	PageRank, 9

4.5 An Analysis of the Independent Citation Network of Articles

It is often the case that co-authors work in different research groups and cite their previous work or the work of their colleagues not only due to thematic similarity and relevance to the current study but also to increase citation counts. Such situations can artificially inflate the popularity of articles, authors, and institutions. To assess the independent significance of articles, a new network model of independent article citations is proposed. The model represents a citation graph of articles from which citations are excluded if at least one author cites their own work or that of a co-author.

Compared to the initial network, the number of nodes decreased by 1361 and the number of edges by 83,313, as isolated nodes were not taken into account. The structure of the connected components resembled that of a complete network: 38,282 articles formed a single component, while the remaining articles were distributed among 77 components, each consisting of 2 to 5 articles.

The proposed citation network, Independent Citation Network (ICN), is smaller than the original one because after excluding dependent citations, some nodes became isolated and were disregarded. The introduced In-degree index in the initial network corresponds to the In-degree index, In-degree (ICN), in the new network. When ranking articles by this index, no substantial changes were observed in the top 10 articles, although the order changed. The leading article remained in the first place, even though it has many co-authors and its independence share (share of citations by not co-authors to the total number of citations) is 0.4. This is because it serves as a diagnostic criterion and is cited by the majority of studies. Traditional centrality indices also showed no significant differences.

Analysis of the independence share in the original network revealed the following patterns: articles with fewer than ten citations tend to have an index value of either 0 or 1, meaning they are either entirely cited by co-authors or not at all. The most popular articles generally reveal an independence share greater than 0.5, although there are some exceptions. Some articles with relatively high citation counts have an independence share of 0.2 or lower, possibly indicating a large number of co-authors. The correlation between In-degree and In-degree (ICN) is high (0.79) but not equal to 1, which highlights the distinct nature of the proposed index compared to the existing one. The developed index can be applied not only to article citation networks but also to other types of citation networks in a similar manner by grouping citations by affiliation or author.

In conclusion, the new centrality indices BI and PI effectively identify articles that are cited by a large number of other publications, especially those focusing on early diagnostics and clinical studies. Another observation arises from examining the independent citation network. Approximately the same articles appear in the Top 10 of the independent citation network based on the In-degree index, although their ranking order has changed.

References

Aroeira, R., & Castanho, M. (2020). Can citation metrics predict the true impact of scientific papers? *The FEBS Journal, 287*(2), 2440–2448. https://doi.org/10.1111/febs.15255

Fan, F. (2024). The new science of COVID-19: A bibliographic and network analysis *arXiv*:2407.15867 [cs.DL]. doi: 10.48550/arXiv.2407.15867.

Pagani, R., Kovaleski, J., & Resende, L. (2015). Methodi Ordinatio: A proposed methodology to select and rank relevant scientific papers encompassing the impact factor, number of citation, and year of publication. *Scientometrics, 105*, 2109–2135.

Su, W. S., & Chang, C. Y. (2024). Root caries in older adults: A co-citation network analysis. *Gerodontology, 41*(4), 445–450. https://doi.org/10.1111/ger.12745

Chapter 5
An Analysis of Terms

No scientific work is complete without professional terminology. The use of particular terminology determines the topic of interest to the researcher. However, very often publications deal with interdisciplinary research involving the study of various factors. Therefore, a targeted study of the terms that distinguish a thematic group (research direction) makes it possible to obtain a broad understanding of the development of various research directions and determine the prospects for the development of these directions.

This chapter provided examples of methods for various processing of terms found in publication abstracts—from simple statistical methods of terminology analysis to network analysis methods using various centrality indices.

Methods for determining the dynamics of terminology change over the years make it possible to identify the most rapidly developing and fading scientific directions. The analysis of terms found in articles forming communities (clusters) demonstrates how terminology shapes the subject of research. The term co-occurrence method illustrates the research area and allows the identification of narrowly specialized priority research directions. Centrality indices characterize the key research topics.

Thus, in Yazdanjue et al. (2025), VOSviewer was used to visualize author keyword networks. Their analysis revealed notable keywords and new topics in the field of social network anonymization. Using word co-occurrence analysis with SciMAT and constructing alluvial diagrams, the researchers studied thematic developments and their temporal evolution. This comprehensive approach led to the formulation of a new taxonomy classifying existing anonymization methodologies. The study not only reflects the current landscape of social network anonymization but also provides predictive insights into future research trajectories.

In Molontay and Nagy (2020), the authors, along with co-authorship analysis, used keyword analysis, which made it possible to comprehensively study the evolution of network science.

In Fan (2024), bibliometric metrics such as the Hirsch index and g-index, as well as network analysis methods, were used to assess co-authorship, institutional

© The Editor(s) (if applicable) and The Author(s), under exclusive license to Springer Nature Switzerland AG 2026

F. Aleskerov et al., *Bibliometric Analysis by Network Models*, Contemporary Systems Thinking, https://doi.org/10.1007/978-3-032-09171-0_5

and international collaboration, and keyword co-usage. Keyword co-usage networks identified central research themes.

In Lozano et al. (2019), a comprehensive network analysis of keyword co-occurrences in the data envelopment analysis (DEA) literature from 2008 to 2017 was conducted. Using network analysis methods, the authors built a co-usage network in which nodes represent keywords and edges denote their co-occurrence in publications. The analysis revealed several closely connected clusters indicating important research topics. Centrality measures identified bridging keywords that link different thematic areas, suggesting interdisciplinary integration in this field. Temporal analysis demonstrated a shift toward new topics.

In Ashoka and Harinarayana (2022), analyzing research trends in the journal "Scientometrics", the authors analyzed keywords from published articles to identify common and emerging research topics. Bibliometric software tools such as VOSviewer and CiteSpace were used in the study. The results reveal dominant research areas and identify changes in the thematic focus of the journal over a certain period. Using these methodologies and tools, the authors provide insights into the evolution of research topics in the field of scientometrics.

To determine the current research landscape and new trends in various research areas, authors often use comprehensive bibliometric analysis. Thus, Shi and Yih (2024) used VOSviewer, CiteSpace, and Microsoft Excel to identify leading countries, influential affiliations, the most prolific and most cited authors and journals. At the same time, high-frequency keywords made it possible to identify current research directions. A similar bibliometric analysis was conducted in Zhang et al. (2023). In this work high-frequency keywords were also identified, allowing central research themes to be determined. Zhu et al. (2024) used VOSviewer and CiteSpace software to assess global research trends on sleep disorders in Parkinson's disease. Co-citation analysis, co-authorship analysis, and keyword co-usage analysis were carried out. The study highlights the emergence of new keywords for research.

In Liao et al. (2024), high-frequency keyword co-occurrence analysis and cluster analysis were carried out to identify current research hotspots, while keyword analysis was used to identify new trends and future research directions. In Cinpolat and Akar (2023), keyword co-occurrence analysis highlighted "Parkinson's disease", "biomarker", and "alpha-synuclein" as known terms, emphasizing their central role in Parkinson's disease biomarker research. In Lin et al. (2022), the words co-occurrences analysis is also performed.

In Utegenova et al. (2025), using keyword analysis, new topics and shifts in research focus over time were identified.

Thus, terminology analysis is an important part of bibliometric analysis.

In this chapter, using data of the Microsoft Academic database of the publications on Parkinson's disease, several options for terminology analysis are considered.

The set of terms was obtained from the abstracts of downloaded publications on Parkinson's disease (PD). The term system included terms, abbreviations of terms, and synonym terms.

The term analysis consisted of:

Table 5.1 Research directions

№	Name of the research area	Number of terms	Number of mentions
1	Biochemistry + cell + genetics (BCG)	139	97,999
2	Medications	147	13,913
3	Symptoms	55	50,671
4	Central nervous system (CNS)	39	47,458
5	Diagnosis	23	20,098
6	Research methods	20	15,939

- Calculating the frequency of term occurrences in the abstracts of the selected publications (by year, total for the entire period);
- Grouping terms by research directions and analyzing frequency of occurrence within each direction (by year, total for the entire period);
- Grouping terms into pairs and triplets and analyzing these combinations, both in the total volume of terms and separately by research directions, including examining the occurrence of pairs and triplets across different directions and time periods;
- Analyzing the centrality of publications related to research directions based on the terms used;
- Analyzing term networks constructed for each research direction.

During the filtering of publications (without DOI, without references, etc.—see Chap. 3), a list of 39,811 articles remained, in which 3,361 terms were identified.

Among these:

- 140 terms had a frequency from 1,000 to 11,179;
- 136 terms had a frequency from 500 to 1,000;
- 849 terms had a frequency from 100 to 500;
- 2,236 terms had a frequency from 1 to 100.

Since our study focused on publications dedicated to Parkinson's disease, the term "parkinson" appeared in them 39,016 times. This term was excluded from the analysis.

From the overall set of terms, groups of terms related to different areas of Parkinson's disease research were identified through expert analysis. Different spellings of the same term (singular and plural forms, abbreviations, acronyms, and synonyms) were considered to be a single term. We grouped terms related to the following directions: biochemistry-cell-genetics (BCG), medications (pharmacology), symptom, central nervous system (CNS), diagnosis (various forms of PD), and clinical research methods (instrumental, laboratory, and other forms of patient examination). Table 5.1 presents the number of terms in each research direction and their frequency of occurrence.

It should be noted that we considered a case where a term was mentioned in the abstract at least once in any form (word, abbreviation, etc.). Even if a term was mentioned multiple times in a single article abstract, only one mention per article was counted. Thus, each term was assigned one article in which it appeared.

5.1 Statistical Analysis of Term Groups Related to Different Research Directions in Parkinson's Disease

5.1.1 An Analysis of Terms in the Overall Volume of Publications

BCG (Biochemistry, Cell, Genetics)

In the overall list of publications, 27,072 articles with terms from the BCG research direction were identified. These articles compose 68% of all downloaded publications. In more than 1,000 publications, 23 terms from this group were used. The most frequently mentioned terms were "dopamine" (10,882 articles), "protein" (8,323 articles), "synuclein" (5,283 articles), as well as "gene" (4,290 articles) and "molecule" (3,835 articles). This group also included the term related to the lrrk2 gene (1,191 publications), mutation of which leads to PD. The fewest publications contained the terms "erk2" (Extracellular Signal Regulated Kinase 2) (6 articles) and "alphasynuclein" (3 articles).

Medications

Terms in the field of Medications (pharmacology) were identified in 9,182 articles (23% of all publications). Ten terms from this group appeared only once during the entire period under review, i.e., in one article. Two terms appeared in the largest number of publications—the drug "levodopa" was mentioned in 2,576 articles, and "neuroprotective" in 2,779 publications.

Symptoms

Terms from the Symptoms group were found in 26,802 publications (67% of all publications). Of these, 14 terms appeared in more than 1,000 publications. Along with terms describing PD symptoms ("tremor"—2,083 articles; "dyskinesia"—1,500 articles; "rigidity"—1,054 articles), the group frequently included the terms "depressed" (2,389 publications) and "sclerosis" (2,087 publications). The least frequent terms were "hyponatraemia"—1 article and "hepatolenticular"—2 articles.

CNS (Central Nervous System)

Terms in this research direction were mentioned in 17,663 publications (44.3% of publications). Of these, 10 terms appeared more than 1,000 times, and 9 terms less than 100 times. The most frequent terms in the publications were “nigra”—4,290 articles and “striatum”—3,527 articles; the least frequent term was “extranigral”—15 publications.

Diagnosis

Terms from the Diagnosis group (recall that the term “parkinson” was not considered) were found in 10,839 publications over 7 years (27% of publications). More than 1,000 occurrences were recorded for 4 terms. The diagnosis “Alzheimer” appeared in 5,539 publications, which accounted for 51.1% of all publications containing terms from this group. Next in frequency (23%) was the term “Lewy” (Dementia with Lewy bodies) (2,486 publications), followed by the terms “huntington” (Huntington’s disease) (1,382 publications) and “amyotrophic” (1,371 publications). Six terms appeared fewer than 100 times. The terms “guam” (Guam disease) and “lethargica” were mentioned in 3 articles each.

Research Methods

Terms in this research direction were mentioned in 9,580 publications over 7 years (24% of publications). Five terms appeared more than 1,000 times. The term “stimulation” in 3,661 articles, “dbs” (deep brain stimulation)—in 2,009, and “mri” (Magnetic resonance imaging) in 1,368 articles. Four terms appeared less than 100 times. The term “tdcs” (Transcranial direct current stimulation) occurred in 99 publications, “tcs” (transcranial sonography)—74 publications, “thalamotomy”—60 publications, and “cag” (coronary angiogram) was mentioned in 50 publications.

The analysis of the number of publications by year showed that there were no sharp changes in the proportion of articles mentioning terms from different fields. Table 5.2 presents the number of publications in each year (from 2015 to 2021) in which at least one term from the considered research area appeared. The table also shows the relative frequency of terms from the given group for each considered year. Note, that an article can contain several terms.

Let us note the general growth in the number of publications over the considered period and, accordingly, the growth in the number of publications mentioning terms. However, the analysis shows that over this 7-year period, no significant changes in research directions were identified. Only a certain tendency towards an increase in works related to diagnostics and a decrease in the number of works related to drugs can be noted.

The quantity and frequency of term mentions determine the interest in a given research area and the direction within the research area. To identify the dynamics

Table 5.2 Number of articles per year with terms

		2015	2016	2017	2018	2019	2020	2021
Number of publications	Total 39,811	4613	4908	5396	5139	5876	6816	7063
BCG	articles	3166	3318	3618	3538	4016	4680	4736
	% to total	68.6	67.6	67.0	68.8	68.3	68.7	67.1
Medications	articles	1136	1212	1254	1187	1323	1490	1580
	% to total	24.6	24.7	23.2	23.1	22.5	21.9	22.4
Symptoms	articles	3129	3287	3598	3493	3912	4604	4779
	% to total	67.8	67.0	66.7	68.0	66.6	67.5	67.7
CNS	articles	2114	2203	2432	2294	2645	2918	3057
	% to total	45.8	44.9	45.1	44.6	45.0	42.8	43.3
Diagnosis	articles	1204	1275	1428	1363	1662	1899	2008
	% to total	26.1	25.9	26.4	26.5	28.2	27.8	28.4
Research methods	articles	1141	1189	1300	1260	1417	1601	1672
	% to total	24.7	24.2	24.1	24.5	24.1	23.5	23.7

of changes within the research area, a temporal analysis of term usage frequency was conducted, with terms grouped into categories. For each term within the group, the number of articles in which it is mentioned per year was counted, as well as the relative frequency of the term within the group. In Fig. 5.1a,b,c,d,e,f, histograms of the relative frequency of term mentions for each year are shown. The histograms illustrate the frequency for the Top-3 terms of each group. The Top-3 includes terms that are mentioned more frequently in articles using terms from a given group than other terms in the same group.

This method of analysis makes it possible to identify trends in changes in researchers' interests in specific local studies within scientific fields.

5.1.2 An Analysis of Terms in Highly Cited Publications

Chapter 4 described publications in various journals with the highest number of citations. In this section, the list of publications from 39,811 articles was ranked by the number of citations from highest to lowest. A comparative analysis was then conducted of the number of terms from each research area found in the Top-50 articles (more than 20 journals) and in the Top-1000 articles. Table 5.3 shows the number of terms in the most cited articles for each research direction.

As can be seen from Table 5.3, in the Top-50 most cited articles, the absolute majority of articles contain terms from the BCG group, while a relative majority remains for articles with terms from the CNS group. Let us note also that the number of articles with terms from the Medications group included in the Top-50 has sharply decreased.

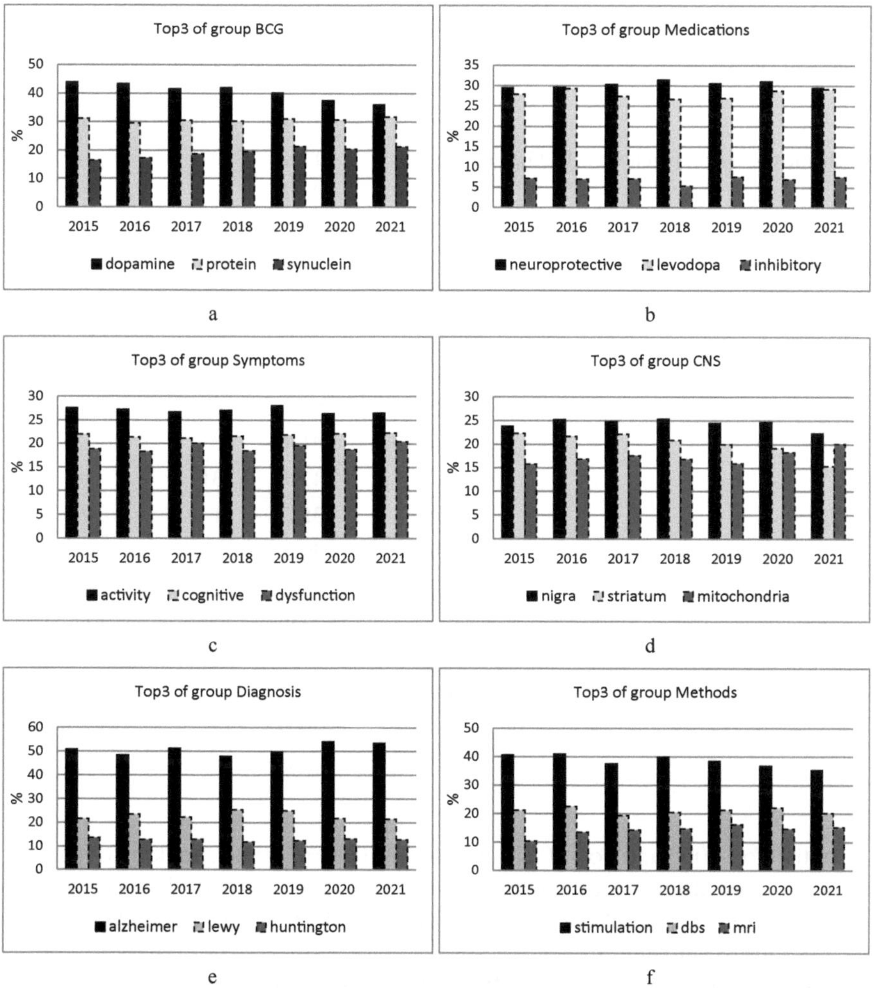

Fig. 5.1 Diagrams of the relative frequency of term occurrence in publications mentioning terms from a given group

Table 5.3 Number of terms in top publications

Research area	Number of terms		
	Total	Top-1000 articles	Top-50 articles
Biochemistry + cell + genetics (BCG)	139	119	44 (31.7%)
Medications	147	54	14 (9.5%)
Symptoms	55	39	17 (30.9%)
Central nervous system (CNS)	39	36	17 (43.6%)
Diagnosis	23	21	9 (39.1%)
Research methods	20	17	6 (30.0%)

Table 5.4 Frequency of term occurrence in top publications

Research area	Term	Number of publications with the term		
		Total	In Top-1000	In top-50
Biochemistry + cell + genetics (BCG)	dopamine	10,882	309	16
	protein	8323	230	20
	synuclein	5283	306	23
Medications	neuroprotective	2779	42	4
	levodopa	2576	74	5
	inhibitory	642	5	0
Symptoms	activity	7236	158	6
	cognitive	5856	165	9
	dysfunction	5166	153	13
CNS	nigra	4290	131	10
	striatum	3527	71	0
	mitochondria	3089	107	7
Diagnosis	alzheimer	5539	59	4
	lewy	2486	159	12
	huntington	1382	9	0
Research methods	stimulation	3661	29	2
	dbs	2009	17	2
	mri	1368	14	0

For the Top-1000 and Top-50 articles in each research area, a ranking of terms by frequency of occurrence was constructed and a comparative analysis was conducted with the overall volume of publications. Table 5.4 illustrates the obtained results for the Top-3 terms shown in Fig. 5.1.

As seen from the table, the most frequently encountered term in the total number of publications does not always maintain its position in highly cited publications (Top-50). It can be assumed that the leading terms in highly cited publications define the direction of research in the given area.

5.2 An Analysis of Terms in Communities

As described above (see Chap. 3), clustering of the article citation graph was carried out to identify homogeneous publications using the Leiden algorithm. In network analysis terminology, researchers refer to clustering of network data as community detection. The concept of a community is related to the classification of objects and, depending on context, it can be equivalent to a class, group, cluster, etc. In this context, clusters represent communities. For these communities, an analysis was conducted of the terms found in the articles forming the communities.

The article citation graph was divided into 117 communities, of which contained more than 30 articles and had an average citation rate of at least 4. Table 5.5 presents the indicators of the 17 selected communities. In addition to the average citation rate, the table also includes another parameter—community density. Community density determines the maximum number of pairwise adjacent graph nodes, i.e., it shows how tightly the nodes are connected to each other. If all nodes are connected to each other, the density is 1; if none are connected to each other, the density is 0. Communities are distinct, in total in all communities there are 39,553 articles.

In each community, the number of articles including terms from the identified research topics was counted. The assignment of an article to a topic was determined by the presence of a term in the abstract; one article could belong to several topics.

Table 5.6 shows the number of terms from each direction found in the 17 identified publication communities, and Table 5.7 shows the number of mentions of terms from each direction in the 17 communities.

As can be seen from Table 5.7, in each community one or another group of terms predominates. As an example, we present the results of the analysis of some of the 17 identified publication communities.

Table 5.5 Indicators of 17 communities identified in the publication citation graph

№	Number of articles	Average citation	Community density	Number of articles with at least 4 citations
1	6136	6.758	0.00075	2370
2	4638	6.827	0.0012	1804
3	4289	10.613	0.00183	1906
4	3805	10.110	0.00177	1631
5	3586	8.155	0.00159	1535
6	3209	7.122	0.00146	1245
7	3076	6.343	0.00163	1090
8	2598	8.611	0.00189	982
9	2356	6.463	0.0018	884
10	2320	9.583	0.00278	881
11	1094	6.255	0.00376	386
12	955	6.789	0.00472	379
13	498	7.889	0.00933	197
14	467	4.738	0.0071	156
15	346	5.014	0.0102	107
16	145	4.551	0.0214	46
17	35	4.885	0.1067	14

Table 5.6 Number of terms in each community

№	Number of terms in each direction					
	BCG	Medications	Symptoms	CNS	Diagnosis	Research methods
1	74	121	43	36	26	15
2	63	49	45	36	23	15
3	73	74	41	37	24	15
4	75	80	39	38	26	14
5	72	71	46	38	25	17
6	73	108	48	37	26	15
7	59	46	46	38	23	16
8	69	69	58	35	23	16
9	65	68	47	34	25	15
10	72	82	41	35	24	16
11	67	39	30	32	19	12
12	66	33	31	35	21	11
13	63	39	35	30	20	12
14	36	19	31	26	16	13
15	33	13	27	28	23	13
16	50	16	22	16	16	7
17	20	1	9	15	4	4

Community No. 1 includes the largest number of publications (6136). Term analysis shows that this community is formed from publications relating to various fields of research. At the same time, it should be noted that the groups BCG and CNS predominate in the publications, since terms from these topics occur more frequently. Table 5.8 presents the number of publications in community No. 1 whose abstracts contain terms included in the Top-10.

In contrast to community No. 1, in community No. 2 (4,638 articles), publications describing symptoms predominate, i.e., this community includes publications describing the symptoms of PD and methods for their investigation. Table 5.9 presents the Top-10 terms from the abstracts of articles in community No. 2.

The two communities with the highest average citation rate (see Table 5.5) are community No. 3 (4,289 articles, average citation 10.61) and community No. 4 (3,805 articles, average citation 10.11). They mainly consist of publications of the topic "Biochemistry, Cell, Genetics (BCG)." More than 50% of the Top-10 terms are from the BCG research field. Table 5.10 presents the Top-10 terms from the abstracts of articles in community No. 3 (a) and community No. 4 (b).

Let us consider community No. 8 (2,598 articles), which included three publications from the Top-10 (by In-degree index). These include the article ranked 1st with the highest number of citations—1,563 ("MDS clinical diagnostic criteria for Parkinson's disease"). Also included are publications ranked 4th ("MDS research

Table 5.7 Number of mentions of terms from each category in the 17 communities

№	BCG	Medications	Symptoms	CNS	Diagnosis	Research methods
1	20,866	3815	5874	9296	3518	524
2	965	441	6626	1671	654	1408
3	14,935	784	4248	4028	2143	317
4	13,335	822	2837	3927	3374	186
5	2883	469	5544	6513	2877	2258
6	4968	3029	4138	2697	734	983
7	2548	623	4641	6795	340	5409
8	1728	573	4174	2195	1127	991
9	1237	713	2678	1557	520	850
10	3536	619	2102	1570	1068	374
11	2247	296	642	1858	362	461
12	3211	224	740	924	680	57
13	1201	414	470	515	291	51
14	136	86	579	174	65	256
15	162	52	411	348	188	180
16	240	28	151	71	148	13
17	37	16	24	73	14	10

Table 5.8 Top-10 terms from the abstracts of articles in community No. 1

Term	Number of articles	Research area
protein	2181	BCG
brain	2044	CNS
neuroprotective	1795	Medications
oxidative	1657	BCG
alzheimer	1639	Diagnosis
nigra	1502	CNS
activity	1343	Symptoms
dopamine	1277	BCG
mitochondria	1089	CNS
microglia	1018	BCG

criteria for prodromal Parkinson's disease") and 6th ("Non-motor features of Parkinson disease"). Note that four publications from the Top-10 (by In-degree index) are in community No. 10 (to be discussed below), and one publication each is in communities 3, 4, and 6.

The main terms in the articles included in community No. 8 relate to the topic Symptoms. The publications present clinical diagnostic criteria for Parkinson's disease (PD). The large number of references to these publications is explained by

Table 5.9 Top-10 terms from the abstracts of articles in community No. 2

Term	Number of articles	Research area
gait	1612	Symptoms
movement	840	Symptoms
cognitive	825	Symptoms
activity	561	Symptoms
postural	559	Symptoms
updrs	557	Research methods
brain	486	CNS
tremor	331	Symptoms
stimulation	246	Research methods
dysfunction	242	Symptoms

Table 5.10 Top-10 terms from the abstracts of articles in community No. 3 and community No. 4

Community No. 3			Community No. 4		
Term	Number of articles	Research area	Term	Number of articles	Research area
protein	1858	BCG	synuclein	2698	BCG
gene	1780	BCG	protein	2381	BCG
genetic	1470	BCG	brain	1118	CNS
mitochondria	1177	CNS	lewy	950	Diagnosis
activity	973	Symptoms	alzheimer	873	Diagnosis
brain	834	CNS	amyloid	759	BCG
cellular	780	BCG	molecular	745	BCG
dysfunction	780	Symptoms	cellular	631	BCG
synuclein	656	BCG	nigra	509	CNS
molecular	647	BCG	activity	508	Symptoms

Note that in the Top-10 of community No. 4, in addition to the terms BCG, CNS, and Symptoms, there are terms from the Diagnosis group. It can be assumed that this community brings together works that study not only of PD, but also of other neurodegenerative diseases

the recommendations of the International Parkinson and Movement Disorder Society (MDS) to use these diagnostic criteria in clinical research and for clinical diagnosis. Table 5.11 presents the Top-10 terms of this community.

Community No. 10 (2,320 articles), average citation—9.58 (third place), maximum number of citations—500. As mentioned above, this community includes 4 articles from the Top-10 by In-degree index—articles ranked 2nd, 3rd, 5th, and 10th (see Table 4.4). The main subject areas of the articles in this community (more than 50% of the articles) include the BCG and Symptoms sections. Table 5.12 shows the Top-10 terms of this community.

Table 5.11 Top-10 terms from the abstracts of articles in community No. 8

Term	Number of articles	Research area
movement	628	Symptoms
cognitive	560	Symptoms
dysfunction	558	Symptoms
dementia	406	Symptoms
brain	400	CNS
rbd	335	CNS
dopamine	292	BCG
tremor	257	Symptoms
lewy	257	Diagnosis
updrs	240	Research methods

Table 5.12 Top-10 terms from the abstracts of articles in community No. 10

Term	Number of articles	Research area
brain	640	CNS
alzheimer	407	Diagnosis
dysfunction	323	Symptoms
protein	320	BCG
synuclein	314	BCG
dopamine	235	BCG
cognitive	230	Symptoms
genetic	195	BCG
sclerosis	190	Symptoms
gene	186	BCG

Analysis of terms from the 17 identified communities showed the subject matter of the publications grouped into each community. Various research areas were combined into one community.

We identified:

- BCG + Symptoms + CNS – 4 communities
- Symptoms – 4 communities
- Symptoms + CNS – 2 communities
- BCG + Medications + Symptoms – 2 communities
- BCG + Diagnosis + CNS – 1 community
- BCG + Symptoms + Research Methods + CNS – 1 community
- BCG + Symptoms – 1 community
- BCG + CNS – 1 community
- BCG + Symptoms + Diagnosis – 1 community

The examples provided show that such clustering is useful for identifying homogeneous communities in a large volume of diverse publications on one global topic. The subject of our analysis was publications on various aspects of Parkinson's disease research. Identifying communities helps determine the level of interest among specialists in a particular research topic, find relevant themes, and track their development.

5.3 An Analysis of Term Combinations

In the abstracts of scientific articles, terms usually relate to general scientific terminology, domain-specific vocabulary, and highly specialized lexicon. In order to more precisely judge the research topic and to identify and analyze the highly specialized topics in a given research area, we conducted an analysis of pairs and triplets of terms. Combinations of two or three terms were constructed by iterating through all the terms within each identified research field and combinations of terms from different research fields.

5.3.1 Pairs and Triplets of Terms Within Research Directions

We perform the analysis of the occurrence of the pairs and triplets of the terms of different topic groups. The pairs or triplets of terms are counted as co-occurrence if the two (three) terms from the two or three groups are mentioned in the abstract of the article simultaneously. And for single article such co- occurrence is counted only once, despite the number of such mentions in the abstract of the article.

Out of the 39,811 articles considered in each direction, only those articles were taken into account that contained pairs or triplets of topics in a given direction.

Group BCG (Included 139 Terms with a Mention Frequency of 97,999 in 27,072 Publications)

a. Analysis of Term Pairs

A total of 6,627 paired combinations of terms were considered, which occurred in 20,192 publications.

Of these, 1,283 pairs occurred once in the articles; 739 pairs—twice. The pair "protein_synuclein" appeared in 3,051 publications. Fifteen pairs occurred in more than 1,000 publications. The Top-3 most frequent terms in this group were: "dopamine"—10,882; "protein"—8,323; "synuclein"—5,283. The relative percentage of pairs with these terms in publications related to the BCG area showed that pairs with the term "dopamine" appeared in 67.5% of publications. Pairs with the term "synuclein" in 67.5%, and with the term "protein" in 87.1% of publications. As an example, Table 5.13 shows the number of publications and the relative frequency of occurrence of some pairs of these terms.

As can be seen from Table 5.13, the most frequently occurring pairs are composed of profile-specific terms. To trace the change in researchers' interest in certain highly specialized studies, an analysis of the number of publications including pairs with highly specialized terminology has been conducted. Here are some examples.

The term "lrrk2"—a gene whose mutation leads to Parkinson's disease—appears in more than 1,000 publications. The number of publications mentioning this term increased over 2015–2021. For example, in 2015, 132 articles with this term were published; in 2019—176; in 2020—231; and in 2021—212 articles. A total of 1,191 articles were published over the entire considered time period. An analysis was carried out on term pairs consisting of the term "lrrk2" and another term from the BCG group. For example, the pair with the highly specialized term "G2019S", defining a mutation of the lrrk2 gene, appeared in 393 publications; with the term "gba" (gene) in 154 publications. At the same time, with profile-specific terms such as "gene"—489 publications, "protein"—510 publications, and "molecule"—198 publications.

The combination with profile-specific terms defines the research area, and with narrowly specialized ones—the specific task within that area. The analysis of combinations of three terms allows for a clearer understanding of the content of the publication.

b. Analysis of Term Triplets

For the group of BCG terms, 67,366 triplets were compiled, which occurred in 15,551 publications. 398 triplets appeared from 100 to 972 publications. For example: the three terms "dopamine_protein_synuclein" appeared in 972 publications, and the triplet "dopamine_gene_protein"—in 541 publications. A large number of publications include triplets consisting of profile-specific terms. Meanwhile, term triplets with two or three narrowly specialized terms, as a rule, occurred much less frequently.

Table 5.13 Number of publications and relative frequency of occurrence of pair combinations with Top-3 most frequent terms

Pair of terms	Number of articles	Frequency of occurrences %
dopamine_protein	2700	13.4
dopamine_synuclein	1840	9.1
dopamine_oxidation	1359	6.7
dopamine_molecule	1282	6.3
protein_synuclein	3051	15.1
molecule_synuclein	1052	5.2
molecule_protein	1898	9.4
cellular_protein	1557	7.7
oxidation_protein	1291	6.4

Thus, 31,511 triplets appeared only once, and 11,004 triplets—in two publications each.

Above, using the example of the lrrk2 gene, we considered the number of publications where the term pair appears. Here are examples of the frequency of occurrence of term triplets, one of which is lrrk2. Most often in article abstracts, two profile-specific and one narrowly specialized term appear simultaneously. For example, the combination "gene_genetic_lrrk2" appeared in 236 publications, and the combination "gene_lrrk2_protein" in 197 publications. However, the greatest interest was in publications where several narrowly specialized terms simultaneously appeared in the abstracts. This allows for a more accurate determination of the nature of the research presented in the article. For example, the number of publications in which the term triplets appeared:

"g2019s_gene_lrrk2" – 162;
"lrrk2_phosphorylation_protein" – 121;
"lrrk2_pink1_snca" – 74;
"lrrk2_snca_synuclein" – 50;
"atp13a2_lrrk2_pink1" – 29.

There were more than 10,000 publications where one article contained a combination of three narrowly specialized terms, such as "a30p_e3_lrrk2" or "cas9_caspase_lrrk2".

A similar analysis was conducted for all groups of terms. We will briefly touch upon each research area.

Group of Terms Medications (147 Terms, Frequency of Mention 13,913 in 9,182 Publications)

a. Analysis of Term Pairs

A total of 1,147 term pair combinations were reviewed. 619 pairs appeared in one publication; 192 pairs in two articles. The medicinal product "levodopa" was mentioned in 2,576 articles. In total, 103 terms were found in pairs with the term "levodopa". Most often, the term "carbidopa" is mentioned in conjunction with this term, since patients with PD are usually prescribed a combination of these drugs. Simultaneously, the terms "carbidopa_levodopa" appear in 370 publications, which is 15.2% of the total number of publications in which terms from this group are found. Next in frequency are "antiparkinson_levodopa"—125 publications (5.1%) and "neuroprotective_superoxide"—124 publications (5.1%).

b. Analysis of Term Triplets

From the terms, 1,609 triplets were composed, which were found in 611 publications. Three terms "carbidopa_entacapone_levodopa" appeared simultaneously in 32 articles. Two term triplets appeared more than 20 times: "neuroprotective_peroxidase_ superoxide" appeared in 27 articles; "pramipexole_ropinirole_rotigotine" in 26 articles. If we count how many articles contain exactly 12 different triples at the same time, then there will be from 10 to 18 articles for different combinations of terms. Exactly one triplet was identified in 1,163 abstracts and exactly two triplets in 223 articles.

Group of Terms Symptoms (55 Terms with a Frequency of Mention 50,671 in 26,802 Publications)

a. Analysis of Term Pairs

From the abstracts of publications including terms from this group, 880 term pair combinations were formed. Three pairs appeared in more than 1,000 publications: "cognitive_dementia" in 1,586 publications (11%); "cognitive_dysfunction"—1,255 (8.7%) and "activity_dysfunction"—1,027 publications (7.1%). 147 pairs were found in one publication. 87 term pairs appeared in two publications, and 108 pairs appeared in at least 100 publications. Overall, pair combinations of this group were found in 14,427 publications. Among the terms of the main symptoms, the pair "rigidity_tremor" was the most frequent—686 publications. The pair "bradykinesia_tremor" was identified in 572 publications, and "bradykinesia_rigidity"—in 563 publications.

b. Analysis of Term Triplets

The analysis included 4,184 term triplets, which were found in 6,710 publications. 1,642 combinations of three terms occurred once, and 662 triplets—twice. The largest number of publications (444) included the triplet "bradykinesia_rigidity_tremor" with the main symptoms of the disease. Another 46 three-term combinations appeared in at least 100 publications. For example, the combination "cognitive_dementia_dysfunction"—373 articles; "cognitive_dementia_movement"—249 publications. Analysis of the 47 most frequently occurring term triplets showed that in 23 combinations one of the terms is "cognitive", and 12 triplets contain the term "movement". Thus, a hypothesis can be proposed regarding the thematic focus of the research.

Group of Terms CNS (39 Terms with a Frequency of Mention 47,458 in 17,663 Publications)

a. Analysis of Term Pairs

A total of 590 term pair combinations were reviewed, found in 8,336 publications. The term "brain" was excluded from consideration as it appeared in 12,253 publications and, accordingly, produced the highest number of pair combinations. The pair "basal_ganglia" was found in the greatest number of publications—1,430 (17%), followed by the pair "nigra_striatum"—1,361 publications (16.3%), and in third place "mitochondria_nigra"—639 publications (7.7%). At least 100 publications featured 67 pairs from this term group. 77 term pairs were found in one publication, and 56 pairs in two publications.

b. Analysis of Term Triplets

From the set of terms in this group, 2,580 three-term combinations were obtained, found in 4,081 publications. 996 combinations occurred once, and 413—twice. More than 100 times, 21 term triplets were encountered. The leading triplet in this group was the combination "basal_cortex_ganglia", which was found in 517 publications (12.7%). More than 300 times the following triplets were found: "basal_ganglia_striatum"—366 publications (8.9%); "nigra_nigrostriatal_striatum"—332 (8.1%); "basal_ganglia_subthalamic"—320 publications (7.8%).

Group of Terms Diagnosis (23 Terms, (Excluding the Term "Parkinson") with a Frequency of Mention 20,098 in 10,839 Publications)

a. Analysis of Term Pairs

A total of 184 term pair combinations were compiled, which were found in 4,224 publications. The term "parkinson", as previously noted, was excluded from consideration. In this group, the term "alzheimer" was the most frequent. 22 pairs were compiled with this term. The pairs "alzheimer_amyotrophic" and "alzheimer_huntington" were found in 1,041 and 1,002 publications respectively. Another nine pairs including the term "alzheimer" appeared in more than 100 publications. The two terms "guam" and "hemiparkinsonism" in pairs with "alzheimer" appeared only once. The second most frequent term in this group "lewy" appeared in a pair with "alzheimer" in 581 publications. In total, 19 pairs were formed with the term "lewy". The pair "hyperhomocysteinemia_lewy" was found in one publication. Six terms in pairs with the term "lewy" appeared in more than 100 publications (Fig. 5.2).

b. Analysis of Term Triplets

For this group of terms, 426 combinations of three terms were compiled, which were found in 1456 publications. Simultaneously, the three terms "alzheimer_amyotrophic_huntington" appeared in 434 article abstracts. Four triplets of terms occurred more than 100 times (see Table 5.14).

Each of the 426 combinations of three terms occurred exactly once, and 50 triplets occurred twice.

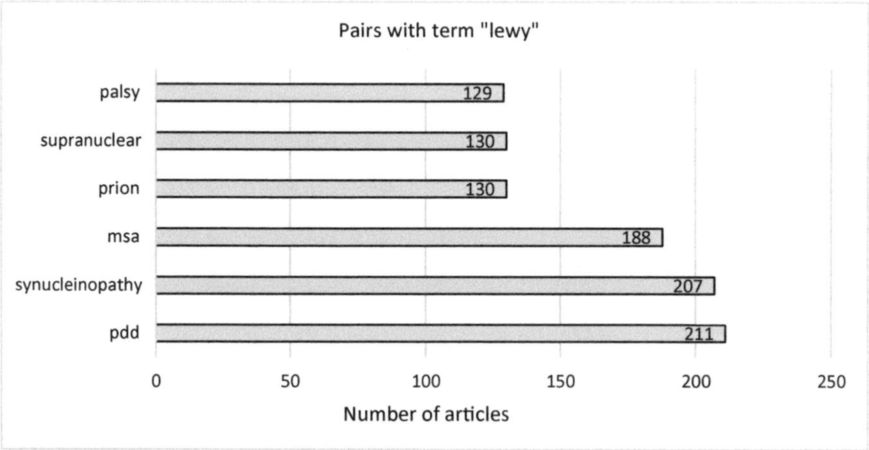

Fig. 5.2 Top-6 pairs with the term "lewy" (41 term pairs had 1 publication each; 10 pairs—two publications each)

Table 5.14 Top-4 combinations of three terms from the diagnosis group

Triplets of terms	Number of articles	Frequency of occurrence %
msa_palsy_supranuclear	185	12.7
lewy_palsy_supranuclear	126	8.7
alzheimer_lewy_pdd	117	8.0
alzheimer_palsy_supranuclear	107	7.3

Research Methods (20 Terms with a Mention Frequency of 15,939 in 9,580 Publications)

a. Analysis of Term Pairs

142 combinations of term pairs in 3,637 publications were considered. 28 pairs appeared in one publication, 12 pairs in two publications. The pair "dbs_stimulation" occurred in 1,930 articles, i.e., in 53.1% of the publications where terms from the Research Methods section appeared; 14 pairs occurred in more than 100 publications. Among them are four pairs with the term "stimulation" and four pairs with the term "dbs" (deep brain stimulation). These pairwise combinations are presented in Table 5.15.

b. Analysis of Term Triplets

249 combinations of three terms were analyzed, which were found in 1233 publications. The most frequent were 5 triplets of terms (Table 5.16).

96 Triplets occurred once and 35 triplets occurred twice.

Table 5.15 Number of publications and relative frequency of occurrence of pairwise combinations with the terms "stimulation" and "dbs"

Pairs of terms	Number of articles	Frequency of occurrence %
dbs_stimulation	1930	53.1
stimulation_surgery	705	19.4
dbs_surgery	603	16.6
stimulation_updrs	401	11.0
dbs_updrs	302	8.3
stimulation_ mri	181	5.0
dbs_mri	135	3.7

Table 5.16 Top-5 combinations of three terms from the research methods group

Triplets of terms	Number of articles	Frequency of occurrence %
dbs_stimulation_surgery	575	46.6
dbs_stimulation_updrs	291	23.6
stimulation_surgery_updrs	133	10.7
dbs_mri_stimulation	132	10.7
dbs_surgery_updrs	126	10.2

The analysis of pairwise combinations revealed the main direction of research related to the application of instrumental methods for the study and treatment of PD. The analysis of combinations of three terms only confirmed this trend.

Thus, the analysis of terms formed by research areas and combinations of term pairs and triplets makes it possible not only to rank research areas by interest in them but also to identify within an area the most relevant and promising research topics for a given time period.

Recently, the number of studies at the intersection of different directions has been growing. To identify the interconnections between different areas of PD research, an analysis of term pairs and triplets from different formed groups was conducted.

5.3.2 *Term Pairs and Triplets from Different Research Directions*

Since we conducted an analysis of scientific publications on Parkinson's disease, it was logical to examine how the Symptom group of terms is related to the groups Medications, Research Methods, Diagnosis, and BCG.

Term Pairs Formed from the Groups Symptoms and Medications

From the two groups, 2,075 pairwise combinations were formed, which were found in 6,449 publications. 854 pairs appeared in the abstract of one publication each, 342 pairs in two publications each, 30 pairs had not less than 100 publications. Most frequently encountered were profile-specific terms from both groups, for example: "activity_neuroprotective"—894 publications or "dysfunction_neuroprotective"—482 publications, as well as pairs consisting of a profile-specific term and a narrowly specialized term, for example, "dyskinesia_levodopa"—754 publications or "activity_levodopa"—400 publications.

The main symptoms of Parkinson's disease are such motor disorders as tremor, rigidity, and bradykinesia.

We examined combinations of these terms with terms from the group Medications. With the profile term "movement", 103 combinations were identified. The term "tremor" was combined with 83, "rigidity" with 70, and "bradykinesia" with 65 terms from the group Medications. Most often, these terms were combined with the drug "levodopa". The highest number of publications was with the pair "tremor_levodopa"—213 articles. The pair "rigidity_levodopa" occurred in 153 publications, and "bradykinesia_levodopa"—in 143 articles. At the same time, there were 115 combinations using the term "cognitive" with terms from the group Medications.

With narrowly specialized terms of cognitive impairments such as "sclerosis"—62, and "dementia"—86 pairs with terms from the group Medications. Most often, the term "sclerosis" was combined with the term "neuroprotective"—173 articles, and the term "dementia" with "levodopa"—127 publications. An analysis was also conducted of drugs included in the group Medications in combination with terms from the group Symptoms. The greatest number of pairs was with the drug "levodopa"—50. Sixteen of these pairs were found in 104 to 754 publications. Four terms in combination with "levodopa" were found in a single publication (amnesia, aphasia, neuralgia, schizoaffective). In the group Medications, 9 terms were found in 1 publication. In pairwise combinations, these terms were found with the following terms from the group Symptoms (Table 5.17).

Table 5.17 Term pairs found in a single publication

Medications	Symptoms
avanir	cognitive
kapikacchu	activity
madecassoside	cognitive
nalbuphine	dyskinesia
nefazodone	dyskinesia
pentamidine	activity
pimozide	sclerosis
sarizotan	dyskinesia
vanoxerine	dyskinesia

Thus, pairwise combinations of terms from the groups Medications and Symptoms allowed the assessment of research interest in studying the effects of drugs on different symptoms.

Term Pairs Formed from the Groups Symptoms and Research Methods

The analysis of pairwise combinations of these groups makes it possible to identify research and treatment methods most frequently encountered with particular terms from the group Symptoms. The combinations of terms were considered in 7,301 publications where they were found (out of 39,811). In total, 687 pairwise combinations were formed, of which 131 pairs occurred only once. 71 term pairs were found in 100 or more publications.

In the group Research Methods, the term "stimulation" was found in the greatest number of publications (3,661). In combination with terms from the group Symptoms, the term "stimulation" formed 49 pairs. The term "dbs" (deep brain stimulation) formed 46 pairs in combination with terms from the group Symptoms. Analysis of these pairwise combinations showed that the highest frequency of occurrence was shown by the pairs presented in Table 5.18. In addition to the number of publications, the table shows the relative frequency (%) in relation to the total number of publications with pairs from these groups.

At the same time, the combination of the terms "stimulation" and "dbs" with the term "kinesia" was found in only one publication. Also, the combinations of the term "dbs" with the terms "aphasia", "atonia", "encephalopathy", "hyperexcitability", and "antidyskinetic" were found once each.

Another term from the group Research Methods that was found in more than 1,000 publications is the term "surgery" (1,017 annotations). The term "surgery" formed 45 pairs, and most often combinations were with the terms "activity" and "movement". These pairs were found in 160 publications (2.19%) each.

Table 5.18 Term pairs with the highest number of publications

Pairs of terms	Number of articles	Frequency of occurrence %
stimulation_activity	858	11.75
stimulation_movement	700	9.59
stimulation_cognitive	528	7.23
stimulation_tremor	480	6.57
stimulation_gait	400	5.48
dbs_movement	422	5.78
dbs_activity	395	5.41
dbs_tremor	344	4.71
dbs_cognitive	277	3.79
dbs_gait	176	2.41

The pairs "surgery_cognitive" and "surgery_tremor" were found in 144 (1.97%) publications each. And with each of the terms "aphasia", "atonia", "encephalopathy", "hyperexcitability", "antidyskinetic", and "paresthesia", the term "surgery" was found in one publication. It should be noted that pairs in different term combinations may be found in the same publication. For example, the pairs "surgery_atonia" and "dbs_atonia" were identified in one publication (2021 doi: https://doi.org/10.1016/j.sleep.2020.11.025). In one publication, the pairs "surgery_aphasia" and "dbs_aphasia" were also identified (2020 doi: https://doi.org/10.1016/j.parkreldis.2019.12.002), and the pairs "surgery, encephalopathy", "dbs, encephalopathy" (2019 doi: https://doi.org/10.25259/SNI-125-2019).

Term Pairs Formed from the Groups Diagnosis and Research Methods

A total of 284 pairs were compiled from the groups of terms Diagnosis and Research Methods, which were found in 1,678 publications. Four pairs were found most frequently, more than 100 times. These pairs and the number of publications in which they were found are presented in Table 5.19. 60 pairs were found in 1 publication each.

As shown above (see Sect. 5.1), the terms "alzheimer" and "stimulation", each in its group, topped the list of terms by the number of publications. The term "alzheimer" was found in 5,539, and the term "stimulation" in 3,661 publications. Second after "alzheimer" was the term "lewy"—2,486 publications. This term in most publications was combined with other terms from the group Research Methods.

Table 5.20 presents the most frequently encountered pairwise term combinations.

Table 5.19 Pairs with the term "alzheimer" with the highest number of publications

Pairs of terms	Number of articles	% out of 1678 articles
alzheimer_stimulation	146	8.70
alzheimer_mri	120	7.15
alzheimer_tomography	111	6.62
alzheimer_neuroimaging	109	6.50

Table 5.20 Pairs with the term "lewy" with the highest number of publications

Pairs of terms	Number of articles	% out of 1678 articles
lewy_tomography	89	5.30
lewy_neuroimaging	80	4.77
lewy_mri	65	3.87
lewy_stimulation	49	2.92
lewy_ 123i	49	2.92

Comparing Tables 5.19 and 5.20, it is seen that the term "lewy", unlike the term "alzheimer", in a pair with the term "stimulation" occurs less frequently than with the terms "tomography", "neuroimaging", and "mri" (Magnetic Resonance Imaging—MRI). In 49 publications, the pair "lewy_123i" is also encountered (123i—a radiopharmaceutical for diagnostics in Parkinson's syndrome).

The analysis of paired combinations of two groups Diagnosis and Research Methods made it possible to identify the research methods most often used for a particular disease diagnosis and determine the degree of interest in them among researchers. Thus, the presented examples of paired term combinations from various research areas showed that such a method of analysis allows revealing connections and interactions of different aspects of research on one topic.

Let us consider the analysis of the combination of three terms from different groups.

We analyzed publications on Parkinson's disease; therefore, terms paired with "parkinson" (the term occurred in 39,016 out of the considered 39,811 publications) were excluded from consideration as appearing in the majority of publications. For example, the pair "parkinson_brain" occurred in 12,166 publications (48.3%). However, in the analysis of three-term combinations, the term "parkinson" was retained.

Triplets of Terms Formed from the Groups Diagnosis, CNS, and Medications

A total of 4,327 triplets were formed, which occurred in 6,219 publications. One triplet "parkinson_brain_neuroprotective" was found in 1,133 publications, 2,489 triplets occurred in one publication each. More than 100 publications contained 30 triplets, among which 6 triplets were the most frequent. Figure 5.3 shows the number of publications in which these triplets occurred. As seen from Fig. 5.3, each triplet contains the term "parkinson", since publications on PD were analyzed. Our triplet sample (4,327) included 1,564 triplets with the term "parkinson". The second most frequent in the Diagnosis group was the term "alzheimer", with which 565 triplets were formed. With the term "lewy", 401 triplets were encountered.

Figures 5.4 and 5.5 present the Top-6 triplets by frequency of occurrence containing the term "alzheimer" (Fig. 5.4) and the term "lewy" (Fig. 5.5).

It should be noted that in all three sets of leading triplets, the combination with two terms "brain" and "neuroprotective" ranks first in the number of publications where these terms occur. All terms included in the top triplets occupy leading positions in their groups (see Fig. 5.1):

Diagnosis—"parkinson" (39,016), "alzheimer" (5,539), "lewy" (2,486);
CNS —"nigra" (4,290), "striatum" (3,527), "mitochondria" (3,089);
Medications—"neuroprotective" (2,779), "levodopa" (2,576).

In addition to the similarity of the triplets that became leaders in the number of publications, each term indicating a diagnosis has its differences. Thus, with the term

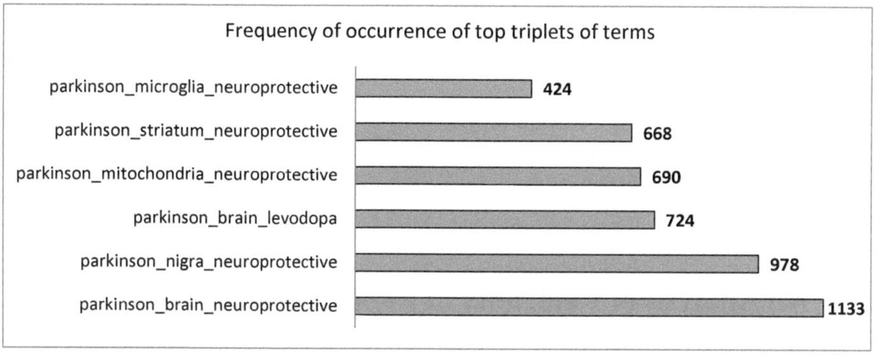

Fig. 5.3 Top-6 combinations of three terms and their frequency of occurrence

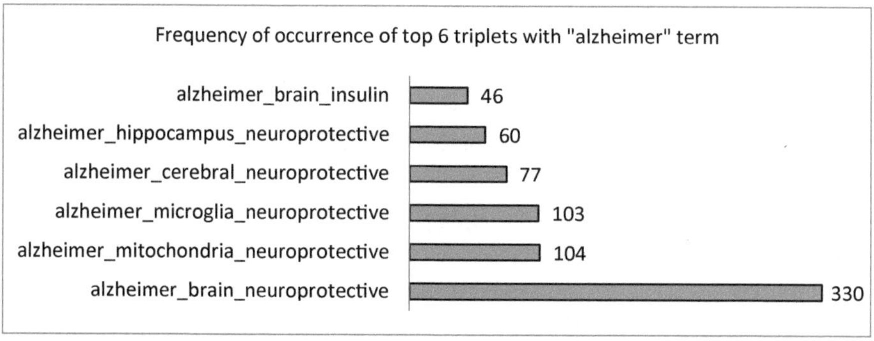

Fig. 5.4 Frequency of occurrence of the Top-6 combinations of three terms containing the term "alzheimer"

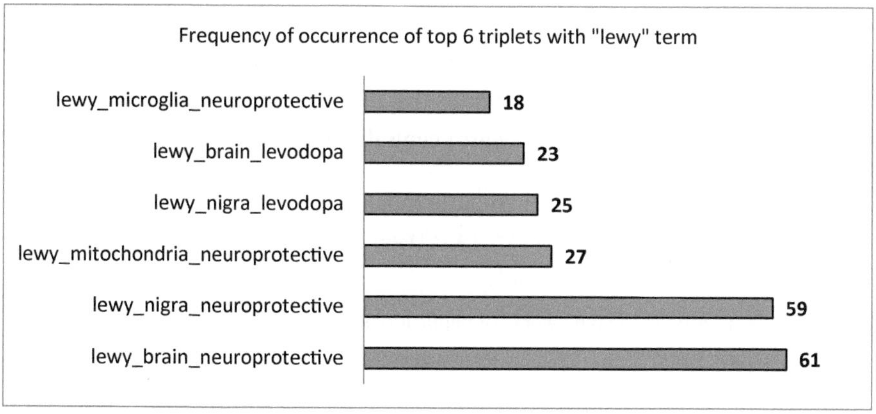

Fig. 5.5 Frequency of occurrence of the Top-6 combinations of three terms containing the term "lewy"

"alzheimer", the Top-6 included triplets "alzheimer_brain_insulin" and "alzheimer_hippocampus_neuroprotective", and with the term "lewy"—"lewy_nigra_levodopa" and "lewy_brain_levodopa".

Thus, the combination of three terms from different research directions expands the researcher's possibilities in searching for specific publications dedicated to narrowly focused directions.

5.4 An Analysis of Centrality of Publications Assigned to Selected Topics

For each identified research direction (topic), including a specific set of terms, publications containing at least one term from the set were selected. A publication could be assigned to several topics if it included terms from different research directions. For each topic, a citation network was constructed consisting only of those articles that were assigned to the respective topic. Isolated nodes were removed from the resulting networks. As a result, six networks corresponding to the 6 topics were obtained:

Biochemistry + Cell + Genetics (BCG)
Medications
Symptoms
Central Nervous System (CNS)
Diagnosis
Research Methods

The main indicators of the publication citation network for each topic are presented in Table 5.21.

For each of the networks, publication centrality indices were calculated. We conducted an analysis by three indices:

- the classical In-degree Index, a degree centrality index, since it shows the number of references to a node;
- the group influence index—BI (Bundle Index);
- the pivotal node influence index—PI (Pivotal Index).

Table 5.21 Indicators for the 6 networks of publications by topics

Name of the indicator	Research area					
	1	2	3	4	5	6
Number of vertices	26,536	9674	8729	26,135	8211	16,928
Number of arcs	192,284	36,214	36,444	158,421	26,513	95,339
Number of connected components	65	122	67	88	82	48
Size of maximal connected component	26,403	9408	8576	25,952	8037	16,826

To analyze the dynamics of index changes in the period 2015–2021, this time range was divided into two periods—2015–2018 and 2019–2021. For each period within one topic, a network was built. Thus, for each of the 6 topics, in addition to the publication citation network for 2015–2021, two more graphs were constructed. Tables 5.22–5.27 present the main indicators of the publication citation network for 2015–2018 (period 1) and 2019–2021 (period 2) for each of the 6 topics. Note that since the article from the 1st period can be cited in an article from the 2nd period, then the article from the first period is included to the network of the 2nd period.

Table 5.22 Indicators of the publication citation network on topic 1 (BCG) 2015–2018 (1) and 2019–2021 (2)

Name of the indicator	Time period	
	1	2
Number of vertices	12,314	24,950
Number of arcs	47,641	144,643
Number of connected components	99	71
Size of maximal connected component	12,103	24,786

Table 5.23 Indicators of the publication citation network on topic 2 (Medications) 2015–2018 (1) and 2019–2021 (2)

Name of the indicator	Time period	
	1	2
Number of vertices	3642	7450
Number of arcs	7332	19,181
Number of connected components	121	84
Size of maximal connected component	3357	7268

Table 5.24 Indicators of the publication citation network on topic 3 (Symptoms) 2015–2018 (1) and 2019–2021 (2)

Name of the indicator	Time period	
	1	2
Number of vertices	11,895	24,467
Number of arcs	40,714	117,707
Number of connected components	111	95
Size of maximal connected component	11,654	24,271

Table 5.25 Indicators of the publication citation network on topic 4 (CNS) 2015–2018 (1) and 2019–2021 (2)

Name of the indicator	Time period	
	1	2
Number of vertices	7914	15,816
Number of arcs	24,324	71,015
Number of connected components	83	52
Size of maximal connected component	7732	15,698

Table 5.26 Indicators of the publication citation network on topic 5 (Diagnosis) 2015–2018 (1) and 2019–2021 (2)

Name of the indicator	Time period	
	1	2
Number of vertices	3915	8748
Number of arcs	8847	27,367
Number of connected components	145	132
Size of maximal connected component	3567	8446

Table 5.27 Indicators of the publication citation network on topic 6 (Research Methods) 2015–2018 (1) and 2019–2021 (2)

Name of the indicator	Time period	
	1	2
Number of vertices	3907	7981
Number of arcs	10,165	26,279
Number of connected components	70	78
Size of maximal connected component	3746	7804

Next, let us consider the values of the three indices under review—In-degree Index, BI (Bundle Index), and PI (Pivotal Index).

5.4.1 *In-Degree Index*

When calculating the indices, articles related to one of the six topics were ranked. Each article was assigned a rank depending on the value of the index. Let us recall that we grouped publications that included terms related to the following research areas (topics):

BCG (biochemistry, cell, genetics);
Medications;
Symptoms;
CNS (central nervous system);
Diagnosis;
Research methods.

Since a single publication could include terms from different topics, it could belong to multiple topics. Table 5.28 presents the values of the In-degree index for the Top-10 publications related to the first topic (BCG). For comparison, the ranks of the BI and PI indices with a quota $q = 1\%$ are also provided for these publications.

These publications also appeared in the Top-10 for other topics. Table 5.29 provides the rank values of these publications for the first topic (BCG) as well as for the other five topics. As seen from Table 5.29, eight out of ten articles contain terms from topic 3 Symptoms, and five publications contain terms from topic 4 CNS. Only one article (No. 3) contains terms from a single topic (BCG).

Table 5.28 Top-10 publications (topic BCG) by In-degree index

(BCG topic) Title of the article	Values	Rank	Values of index for q = 1%	
	In-degree	In-degree	BI	PI
The Roles of PINK1, Parkin and Mitochondrial Fidelity in Parkinson's Disease	0.00232	1	0	0
α-Synuclein strains cause distinct synucleinopathies after local and systemic administration	0.00223	2	0	0
A meta-analysis of genome-wide association studies identifies 17 new Parkinson's disease risk loci	0.00210	3	0	0
Gut Microbiota Regulate Motor Deficits and Neuroinflammation in a Model of Parkinson's Disease	0.00209	4	0	0
Phosphoproteomics reveals that Parkinson's disease kinase LRRK2 regulates a subset of Rab GTPases	0.00204	5	0	0
Identification of novel risk loci, causal insights, and heritable risk for Parkinson's disease: a meta-analysis of genome-wide association studies	0.00188	6	0	0
The epidemiology of Parkinson's disease: risk factors and prevention	0.00176	7	0	0
Epidemiology of Parkinson's disease	0.00174	8	0	0
Gut microbiota are related to Parkinson's disease and clinical phenotype	0.00170	9	0	0
Non-motor features of Parkinson disease	0.00152	10	0.0373	0.1157

The absence of terms from a certain topic in a publication is indicated in the table with a value of 0. Let us recall that the term "parkinson" was excluded from the topic Diagnosis, since the analyzed publications are on Parkinson's disease theme. As can be seen from Table 5.29, the articles included in the Top-10 generally hold fairly high ranks in other research areas as well. This indicates that these publications receive a high number of citations from publications of various topics.

Let us consider the rankings across six research areas for the Top-10 publications on topic No. 2 (Medications), presented in Table 5.30. Publications on this topic have fewer citations, which is reflected in the rank values.

Terms from the topic Symptoms (3) are quite frequent in medical publications. The description of symptoms is essential for differential diagnosis and is often used in drug research. This terminology is also often found in studies of the CNS or brain structures. Table 5.31 presents the Top-10 works in which terms from the topic Symptoms are found. The table also shows the rank values of these publications in the other five topics.

As can be seen from Table 5.31, the articles in this list also hold fairly high ranks in the other five topics.

The analysis of the dynamics of changes in the In-degree index was carried out for two time periods: 2015–2018 and 2019–2021. The works were ranked according to the index value in the respective time period. It should be taken into account

Table 5.29 Top 10 articles on topic 1 (BCG) and their ranks for other topics

Rank	Title of the article (topic 1)	In-degree rank on themes				
		2	3	4	5	6
1	The Roles of PINK1, Parkin and Mitochondrial Fidelity in Parkinson's Disease	0	4	1	0	0
2	α-Synuclein strains cause distinct synucleinopathies after local and systemic administration	0	10	0	9	0
3	A meta-analysis of genome-wide association studies identifies 17 new Parkinson's disease risk loci	0	0	0	0	0
4	Gut Microbiota Regulate Motor Deficits and Neuroinflammation in a Model of Parkinson's Disease	0	7	3	0	0
5	Phosphoproteomics reveals that Parkinson's disease kinase LRRK2 regulates a subset of Rab GTPases	0	8	0	0	0
6	Identification of novel risk loci, causal insights, and heritable risk for Parkinson's disease: a meta-analysis of genome-wide association studies	0	15	19	0	0
7	The epidemiology of Parkinson's disease: risk factors and prevention	1	9	0	0	0
8	Epidemiology of Parkinson's disease	0	5	4	3	0
9	Gut microbiota are related to Parkinson's disease and clinical phenotype	0	0	11	0	0
10	Non-motor features of Parkinson disease	0	3	0	0	0

that publications released at the end of 2018 and later were not included in the network built for the first period (2015–2018). The In-degree index values for the Top-10 publications are presented in Table 5.32 (for 2015–2018) and Table 5.33 (for 2019–2021). The tables also provide the article ranks for the entire period under consideration (2015–2021).

Table 5.33 shows the publication dates. Articles No. 1 and No. 10 were published in 2019 and were not included in the network built for the period 2015–2018. It should be noted that articles published in the second half of 2018 (No. 2, 4, 8) did not have time to accumulate many citations and therefore did not receive high ranks. However, over the entire period 2015–2021, these articles made it into the Top-10 in terms of citations. It is also worth noting that some articles with high ranks in the period 2015–2018 later lost their positions and decreased in rank. The number of citations to these works decreased.

As an illustration, we present the indices of the Top-10 articles on topic 4 (CNS), obtained for the period 2015–2018 (Table 5.34) and for the period 2019–2021

Table 5.30 Top-10 articles on topic 2 (Medications) and their ranks for other topics

In-degree rank	Title on the articles (topic 2)	Ranks by themes				
		1	3	4	5	6
1	The epidemiology of Parkinson's disease: risk factors and prevention	7	9	0	0	0
2	International Parkinson and movement disorder society evidence-based medicine review: Update on treatments for the motor symptoms of Parkinson's disease	65	23	23	0	10
3	Pathophysiology of L-dopa-induced motor and non-motor complications in Parkinson's disease	35	20	0	0	0
4	Levodopa-carbidopa intestinal gel in advanced Parkinson's disease: final 12-month, open-label results	734	95	0	0	45
5	MDS research criteria for prodromal Parkinson's disease	14	2	0	0	0
6	Exenatide once weekly versus placebo in Parkinson's disease: a randomised, double-blind, placebo-controlled trial	45	53	0	0	241
7	EuroInf: A Multicenter Comparative Observational Study of Apomorphine and Levodopa Infusion in Parkinson's Disease	0	76	0	0	18
8	Global long-term study on motor and non-motor symptoms and safety of levodopa-carbidopa intestinal gel in routine care of advanced Parkinson's disease patients; 12-month interim outcomes	0	251	0	0	105
9	Diagnosis and Treatment of Parkinson Disease: A Review	63	28	0	0	51
10	Levodopa therapy for Parkinson's disease: Pharmacokinetics and pharmacodynamics	249	270	760	0	0

(Table 5.35). The tables show the index values and ranks for the full period of consideration 2015–2021.

As with the previously examined networks for topic 1 (BCG), the works published in 2019 (No. 6 and 9 in Table 5.35) were not included in the network for the period 2015–2018. All articles except No. 1, which had high ranks in the period 2015–2018, lost their positions (see Table 5.34), i.e., the number of citations to these works in the network for topic 4 decreased in subsequent years. Thus, in the years following publication, the number of citations to articles increases and then begins to decline.

Table 5.31 Top-10 articles on topic 3 (Symptoms) and their ranks for other topics

In-degree rank	Title of the article (topic 3)	Ranks by themes				
		1	2	4	5	6
1	MDS clinical diagnostic criteria for Parkinson's disease	0	0	0	0	0
2	MDS research criteria for prodromal Parkinson's disease	14	5	0	0	0
3	Non-motor features of Parkinson disease	10	0	0	0	0
4	The Roles of PINK1, Parkin and Mitochondrial Fidelity in Parkinson's Disease	1	0	1	0	0
5	Epidemiology of Parkinson's disease	8	0	4	4	0
6	Cognitive decline in Parkinson disease	33	0	10	5	0
7	Gut Microbiota Regulate Motor Deficits and Neuroinflammation in a Model of Parkinson's Disease	4	0	3	0	0
8	Phosphoproteomics reveals that Parkinson's disease kinase LRRK2 regulates a subset of Rab GTPases	5	0	0	0	0
9	The epidemiology of Parkinson's disease: risk factors and prevention	7	1	0	0	0
10	α-Synuclein strains cause distinct synucleinopathies after local and systemic administration	2	0	0	1	0

This trend is observed not only for networks built on thematic publications but also for the general publication network described in Chap. 4.

However, a number of publications retained positions in the Top-10 during 2015–2021, both in various thematic areas and in the total volume of publications. For example, the article "The Roles of PINK1, Parkin and Mitochondrial Fidelity in Parkinson's Disease" (doi: https://doi.org/10.1016/j.neuron.2014.12.007), published in January 2015 in the journal "Neuron," ranked 1st (highest number of citations) in thematic networks 1 (BCG) and 4 (CNS), was in the Top-10 in network 3 (Symptoms), and held 8th position in the overall publication network Top-10. The article "Epidemiology of Parkinson's disease" (doi: https://doi.org/10.1007/s00702-017-1686-y; "Journal of Neural Transmission"; Publisher Springer Science; 2017 Aug) was in the Top-10 in thematic networks 1, 3, 4, and 5 and ranked 3rd in the overall publication network Top-10.

Table 5.32 Top-10 articles on topic 1 (BCG) in the period 2015–2018. In-degree index value and ranks for the periods 2019–2021 and 2015–2021

Rank	Title of the article (topic 1- BCG)	In-degree value 2015–2018	Rank 2019–2021	Rank 2015–2021
1	The Roles of PINK1, Parkin and Mitochondrial Fidelity in Parkinson's Disease	0.004093	7	1
2	α-Synuclein strains cause distinct synucleinopathies after local and systemic administration	0.004051	9	2
3	Phosphoproteomics reveals that Parkinson's disease kinase LRRK2 regulates a subset of Rab GTPases	0.002918	6	5
4	Gut microbiota are related to Parkinson's disease and clinical phenotype	0.002309	11	9
5	Oxidative stress and Parkinson's disease	0.002204	19	12
6	MDS research criteria for prodromal Parkinson's disease	0.002204	26	14
7	Mitochondrial dysfunction and mitophagy in Parkinson's: from familial to sporadic disease	0.002057	75	27
8	Gut Microbiota Regulate Motor Deficits and Neuroinflammation in a Model of Parkinson's Disease	0.002057	3	4
9	Pathophysiology of L-dopa-induced motor and non-motor complications in Parkinson's disease	0.002036	118	35
10	Neuroinflammation in Parkinson's disease and its potential as therapeutic target	0.001952	20	13

5.4.2 Indices BI (Bundle Index) and PI (Pivotal Index)

Analysis of the BI and PI indices of the thematic networks showed that the Top-10 articles list for each of the six topics coincided. This is mainly because the top articles have far more citations than the others and thus they are on the top of both indices.

Table 5.36 presents the values of BI and PI indices and the In-degree rank for the Top-10 articles on topic 1 (BCG) for the period 2015–2021.

Let us consider the change of indices in dynamics. The analysis of the dynamics of BI and PI indices, as well as the In-degree index, was conducted for two time periods: 2015–2018 and 2019–2021. The works were ranked by the index value in the given time period. Since the Top-10 articles list for each of the six topics coincides for BI

Table 5.33 Top-10 articles on topic 1 (BCG) in the period 2019–2021. In-degree index value and ranks for the periods 2015–2018 and 2015–2021

Rank	Title of the article (topic 1- BCG)	Date of publication	In-degree value 2019–2021	Rank 2015–2018	Rank 2015–2021
1	Identification of novel risk loci, causal insights, and heritable risk for Parkinson's disease: a meta-analysis of genome-wide association studies	2019 Dec	0.002502	–	6
2	A meta-analysis of genome-wide association studies identifies 17 new Parkinson's disease risk loci	2017 Oct	0.002253	17	3
3	Gut Microbiota Regulate Motor Deficits and Neuroinflammation in a Model of Parkinson's Disease	2016 Dec	0.002094	8	4
4	Epidemiology of Parkinson's disease	2017 Aug	0.002080	127	8
5	The epidemiology of Parkinson's disease: risk factors and prevention	2016 Nov	0.001880	27	7
6	Phosphoproteomics reveals that Parkinson's disease kinase LRRK2 regulates a subset of Rab GTPases	2016 Jan	0.001756	9	5
7	The Roles of PINK1, Parkin and Mitochondrial Fidelity in Parkinson's Disease	2015 Jan	0.001742	1	1
8	Non-motor features of Parkinson disease	2017 Jul	0.001707	69	10

(continued)

Table 5.33 (continued)

Rank	Title of the article (topic 1- BCG)	Date of publication	In-degree value 2019–2021	Rank 2015–2018	Rank 2015–2021
9	α-Synuclein strains cause distinct synucleinopathies after local and systemic administration	2015 Jun	0.001631	2	2
10	Transneuronal Propagation of Pathologic α-Synuclein from the Gut to the Brain Models Parkinson's Disease	2019 Aug	0.001534	–	20

and PI indices, we provide as an example in Table 5.37 the value of the BI index for the BCG topic for the period 2015–2018 and the ranks for the periods 2019–2021.

Let us provide another example. Table 5.38 presents the values of BI and PI indices and In-degree rank for the Top-10 articles on topic 4 (CNS) for the period 2015–2021.

The value of the BI index for the period 2015–2018 and ranks for the periods 2019–2021 are presented in Table 5.39.

5.5 An Analysis of Terminology Centrality Indices Across Six Research Directions of Parkinson's Disease

For each of the six research directions, a co-occurrence network of terms in articles was constructed taking into account the centrality of the articles. The network is built as follows.

In the abstracts of articles, at least one pair of terms from the same topic is identified. There may be several such pairs. These terms are the nodes of the network. There is an edge between terms if a pair of terms simultaneously appears (as a pair) in article abstracts. The weight of the undirected edge between the terms (network nodes) is calculated as the sum of the centralities of all articles whose abstract simultaneously includes this pair of terms. Then, for each network, the centrality of the terms is calculated. For the term network built based on article centrality using the In-degree Index, the centrality of terms is calculated using the In-degree index. For the network built based on the Bundle Index (BI), the centrality of terms is calculated using BI. For the term network built based on the Pivotal Index (PI), the centrality

Table 5.34 Top-10 articles on topic 4 (CNS) in the period 2015–2018. In-degree index value and ranks for the periods 2019–2021 and 2015–2021

Rank	Title of the article (topic 4- CNS)	In-degree value 2015–2018	Rank 2019–2021	Rank 2015–2021
1	The Roles of PINK1, Parkin and Mitochondrial Fidelity in Parkinson's Disease	0.0067	1	1
2	Mitochondrial dysfunction and mitophagy in Parkinson's: from familial to sporadic disease	0.0033	29	8
3	Neuroinflammation in Parkinson's disease and its potential as therapeutic target	0.0031	8	2
4	Therapeutic deep brain stimulation reduces cortical phase-amplitude coupling in Parkinson's disease	0.0028	21	9
5	Oxidative stress and Parkinson's disease	0.0023	10	6
6	Parkinson risk in idiopathic REM sleep behavior disorder: Preparing for neuroprotective trials	0.0022	83	25
7	α-Synuclein binds to TOM20 and inhibits mitochondrial protein import in Parkinson's disease	0.0022	13	11
8	New Clinical Subtypes of Parkinson Disease and Their Longitudinal Progression: A Prospective Cohort Comparison With Other Phenotypes	0.0021	34	17
9	Elevated Mitochondrial Bioenergetics and Axonal Arborization Size Are Key Contributors to the Vulnerability of Dopamine Neurons	0.00189	15	13
10	Cognitive impairment and resting-state network connectivity in Parkinson's disease	0.00185	135	40

of terms is calculated using the PI index. For BI and PI indices, quotas of 1%, 5%, and 10% were considered. Group size restrictions $k = 5$ were applied.

Thus, the network allows for identifying terms that not only occur frequently but also appear in the most cited publications. At the same time, the new centrality indices allow for highlighting terms that most frequently appear in narrowly specialized articles actively citing each other. We carried out the calculation of the considered In-degree index, Bundle Influence index (BI), and Pivotal Influence index (PI) for the compiled list of terms for each research direction of PD—BCG (biochemistry,

Table 5.35 Top-10 articles on topic 4 (CNS) in the period 2019–2021. In-degree index value and ranks for the periods 2015–2018 and 2015–2021

Rank	Title of the article (topic 4-CNS)—year of publication	In-degree value 2019–2021	Rank 2015–2018	Rank 2015–2021
1	The Roles of PINK1, Parkin and Mitochondrial Fidelity in Parkinson's Disease 2015 Jan	0.0028	1	1
2	Epidemiology of Parkinson's disease. 2017 Aug	0.0022	219	4
3	Gut Microbiota Regulate Motor Deficits and Neuroinflammation in a Model of Parkinson's Disease 2016 Dec	0.0019	20	3
4	Dopamine oxidation mediates mitochondrial and lysosomal dysfunction in Parkinson's disease 2017 Sep	0.0019	65	7
5	Cognitive decline in Parkinson disease 2017 Apr	0.00179	117	10
6	Transneuronal Propagation of Pathologic α-Synuclein from the Gut to the Brain Models Parkinson's Disease. 2019 Aug	0.00174	–	15
7	Mitochondrial dysfunction in Parkinson's disease. 2015 Jun	0.00173	12	5
8	Neuroinflammation in Parkinson's disease and its potential as therapeutic target 2015	0.00166	3	2
9	Identification of novel risk loci, causal insights, and heritable risk for Parkinson's disease: a meta-analysis of genome-wide association studies 2019 Dec	0.00163	–	19
10	Oxidative stress and Parkinson's disease 2017 Mar	0.0015	5	6

cell, genetics), Medications, Symptom, CNS (central nervous system), Diagnosis, Research methods.

In Sect. 5.1 (Table 5.2), the results of a statistical analysis of the number of articles in which at least one term from the considered research directions was mentioned were provided. A simple calculation allows for identifying terms frequently encountered in publications. Network centrality calculation using the In-degree index enabled identifying terms most frequently found in the most cited articles. This allows for identifying the most significant research areas within the studied directions.

Group of terms BCG (Biochemistry + Cell + Genetics)

Table 5.36 Values of BI and PI indices and In-degree rank for the Top-10 articles on topic 1 (BCG) for the period 2015–2021

Rank	Title of the article (topic 1 BCG)	BI (1%)	PI (1%)	In-degree rank
1	Non-motor features of Parkinson disease	0.0373	0.1157	10
2	Selective neuronal vulnerability in Parkinson disease	0.0347	0.1076	11
3	Oxidative stress and Parkinson's disease	0.0266	0.0826	12
4	Neuroinflammation in Parkinson's disease and its potential as therapeutic target	0.0231	0.0716	13
5	MDS research criteria for prodromal Parkinson's disease	0.0226	0.06993	14
6	Colonic bacterial composition in Parkinson's disease	0.0199	0.0617	15
7	Pathological α-synuclein transmission initiated by binding lymphocyte-activation gene 3	0.0182	0.0564	16
8	Dopamine oxidation mediates mitochondrial and lysosomal dysfunction in Parkinson's disease	0.0173	0.0535	17
9	T cells from patients with Parkinson's disease recognize α-synuclein peptides	0.0166	0.0514	18
10	Transneuronal Propagation of Pathologic α-Synuclein from the Gut to the Brain Models Parkinson's Disease	0.0164	0.0507	19

The main indicators of the term network for the BCG direction are presented in Table 5.40.

As an example, Table 5.41 provides the values of the In-degree index for the Top-10 terms of the BCG group. For comparison, the table shows the ranks of these terms by the BI and PI indices for the quota $q = 1\%$ and $k = 5$. Table 5.42 shows the Top-10 terms by frequency of occurrence in the total number of publications.

The top three terms (Table 5.42) coincide with the three terms listed in Table 5.41; however, the ranks of these terms have changed. Thus, the term "dopamine" is leading in the number of mentions in articles (10,882 articles). The term "protein" ranks second in frequency (8,323 articles), and the term "synuclein" ranks third (5,283 articles). However, in the most cited articles, "protein" and "synuclein" occur more often than the term "dopamine". Here, the leader becomes the term "protein." It can be assumed that articles that study of various proteins, including synuclein, are cited more frequently. It should also be noted that Table 5.41 includes more specific terms—"lysosomal" (a cellular component) and the gene lrrk2.

Table 5.37 Top-10 articles on topic 1 (BCG) for the period 2015–2018. BI index ranks for 2015–2018 and ranks for the periods 2019–2021

Rank	Title of the article (topic 1- BCG)	BI rank (1%) 2015–2018	BI rank (1%) 2019–2021
1	The Roles of PINK1, Parkin and Mitochondrial Fidelity in Parkinson's Disease	1	3
2	α-Synuclein strains cause distinct synucleinopathies after local and systemic administration	2	5
3	Phosphoproteomics reveals that Parkinson's disease kinase LRRK2 regulates a subset of Rab GTPases	3	2
4	Gut microbiota are related to Parkinson's disease and clinical phenotype	4	7
5	Oxidative stress and Parkinson's disease	5	15
6	MDS research criteria for prodromal Parkinson's disease	5	22
7	Mitochondrial dysfunction and mitophagy in Parkinson's: from familial to sporadic disease	6	18
8	Gut Microbiota Regulate Motor Deficits and Neuroinflammation in a Model of Parkinson's Disease	6	138
9	Pathophysiology of L-dopa-induced motor and non-motor complications in Parkinson's disease	7	62
10	Neuroinflammation in Parkinson's disease and its potential as therapeutic target	8	16

The Top-10 terms by the Group Influence Index BI for the quota q = 1% are presented in Table 5.43. For comparison, the table shows the ranks of these terms by the In-degree and PI indices. The Top-10 terms by the Pivotal Influence Index PI (q = 1%) are presented in Table 5.44, which also provides the ranks of these terms by the In-degree and BI indices for the Top-10 terms of the BCG group.

The BI and PI indices showed that in highly specialized articles and articles that actively cite each other, three terms remain leaders: "dopamine", "synuclein", and "protein". However, as can be seen from Tables 5.41–5.44, the Top-10 terms by each index differ slightly. Thus, the terms "lysosomal" and "lrrk2", found in frequently cited articles (9th and 10th rank by the In-degree index), fall into the third and fourth tens by the BI and PI indices.

Let us consider the Top-10 terms by the BI and PI indices for quotas q = 5% and q = 10% for the BCG group. Table 5.45 presents the BI index values, and Table 5.46

Table 5.38 Values of BI and PI indices and In-degree rank for the Top-10 articles on topic 4 (CNS) for the period 2015–2021

Rank	Title of the article (topic 4-CNS)	BI (1%)	PI (1%)	In-degree rank
1	Neuroinflammation in Parkinson's disease and its potential as therapeutic target	0.0505	0.0751	2
2	Gut Microbiota Regulate Motor Deficits and Neuroinflammation in a Model of Parkinson's Disease	0.0370	0.0610	3
3	Epidemiology of Parkinson's disease	0.0351	0.0589	4
4	Mitochondrial dysfunction in Parkinson's disease	0.0316	0.0549	5
5	Oxidative stress and Parkinson's disease	0.03099	0.0542	6
6	Dopamine oxidation mediates mitochondrial and lysosomal dysfunction in Parkinson's disease	0.0293	0.0522	7
7	Mitochondrial dysfunction and mitophagy in Parkinson's: from familial to sporadic disease	0.0272	0.0497	8
8	Therapeutic deep brain stimulation reduces cortical phase-amplitude coupling in Parkinson's disease	0.0257	0.0478	9
9	Cognitive decline in Parkinson disease	0.0228	0.0441	10
10	α-Synuclein binds to TOM20 and inhibits mitochondrial protein import in Parkinson's disease	0.02231	0.0435	11

the PI index values (q = 5%) for these terms. For the quota q = 10%, the index values are given in Table 5.47 for BI and in Table 5.48 for PI. The tables also provide the ranks of these terms by the In-degree index.

Comparing the lists of terms in Tables 5.43–5.48, we can see how the value of the quota affects the composition of the Top-10 terms. Thus, in terms of the number of mentions in articles, three main terms are "dopamine" (10,882 articles), "protein" (8,323 articles), and "synuclein" (5,283 articles). In the most cited articles, "protein" and "synuclein" (1st and 2nd ranks by the In-degree index) appear more frequently than the term "dopamine" (3rd rank by the In-degree index). According to the BI and PI indices for quota q = 1%, the terms "dopamine", "synuclein", and "protein" have ranks 1, 2, and 3, respectively. For quota q = 5%, by the BI index, the term "dopamine" ranks 3rd, "synuclein"—18th, and "protein"—7th. By the PI index value (q = 5%), the term "dopamine" has rank 1, "synuclein"—9th, and "protein"—4th.

Table 5.39 Top-10 articles on topic 4 (CNS) for the period 2015–2018. BI index ranks for 2015–2018 and ranks for the periods 2019–2021

Rank	Title of the article (topic 4-CNS)	BI rank (1%) 2015–2018	BI rank (1%) 2019–2021
1	The Roles of PINK1, Parkin and Mitochondrial Fidelity in Parkinson's Disease	1	1
2	Mitochondrial dysfunction and mitophagy in Parkinson's: from familial to sporadic disease	2	7
3	Neuroinflammation in Parkinson's disease and its potential as therapeutic target	3	8
4	Therapeutic deep brain stimulation reduces cortical phase-amplitude coupling in Parkinson's disease	4	19
5	Oxidative stress and Parkinson's disease	5	10
6	Parkinson risk in idiopathic REM sleep behavior disorder: Preparing for neuroprotective trials	6	48
7	α-Synuclein binds to TOM20 and inhibits mitochondrial protein import in Parkinson's disease	7	13
8	New Clinical Subtypes of Parkinson Disease and Their Longitudinal Progression: A Prospective Cohort Comparison With Other Phenotypes	8	29
9	Elevated Mitochondrial Bioenergetics and Axonal Arborization Size Are Key Contributors to the Vulnerability of Dopamine Neurons	9	15
10	Cognitive impairment and resting-state network connectivity in Parkinson's disease	10	56

Table 5.40 Indicators for the BCG term network

Name of the indicator	Value
Number of vertices	139
Number of arcs	6269
Size of maximal connected component	139
Number of connected components	1
Density	0.654

Table 5.41 In-degree, values

№	Terms	In-degree	Rank	
			BI	PI
1	protein	0.0846	3	3
2	synuclein	0.0758	2	2
3	dopamine	0.0734	1	1
4	gene	0.0444	4	6
5	genetic	0.0385	5	5
6	molecule	0.0342	8	7
7	cellular	0.0294	6	9
8	oxidation	0.0278	7	10
9	lysosomal	0.0195	39	49
10	lrrk2	0.0188	24	39

Table 5.42 Total number of publications

№	Terms	Number of publications
1	dopamine	10,882
2	protein	8323
3	synuclein	5283
4	gene	4290
5	molecule	3835
6	genetic	3493
7	oxidation	2969
8	cellular	2859
9	nucleus	2479
10	microglia	1553

For quota q = 10%, by the BI index, the terms "dopamine", "synuclein", and "protein" have ranks 7th, 1st, and 12th, respectively, and by the PI index—7th, 13th, and 4th, respectively. For quotas q = 5% and 10%, new terms appear in the Top-10 by the BI and PI indices, which for quota q = 1% and the In-degree index were in the second, third, or fourth tens.

Table 5.43 BI index for the quota q = 1%

№	Terms	BI q = 1%	Rank	
			In-degree	PI
1	dopamine	0.0314	3	1
2	synuclein	0.0298	2	2
3	protein	0.0297	1	3
4	gene	0.0289	4	6
5	genetic	0.0262	5	5
6	cellular	0.0245	7	9
7	oxidation	0.0244	8	10
8	molecule	0.0243	6	7
9	enzyme	0.0214	19	12
10	neurotoxin	0.01	25	13

Table 5.44 PI index for the quota q = 1%

№	Terms	PI q = 1%	Rank	
			In-degree	BI
1	dopamine	0.0375	3	1
2	synuclein	0.0362	2	2
3	protein	0.0333	1	3
4	mrna	0.0318	32	15
5	genetic	0.0291	5	5
6	gene	0.0288	4	4
7	molecule	0.0275	6	8
8	membrane	0.0268	14	12
9	cellular	0.0266	7	6
10	oxidation	0.0266	8	7

5.5.1 Term Group Medications

Main indicators of the term network for the direction Medications (Table 5.49).

Table 5.50 presents the values of the In-degree index for the Top-10 terms of the Medications group. For comparison, the table gives the ranks of the terms by BI and PI indices for quota q = 1% and k = 5.

Top-10 terms by the values of BI and PI indices (q = 1%) are presented respectively in Tables 5.51 and 5.52.

Let us compare the lists of terms presented according to the three indices with the list of the Top-10 terms leading in the overall list of publications (Table 5.53).

Table 5.45 BI index for the quota q = 5%

№	Term	BI q = 5%	Rank	
			In-degree	PI
1	oxidation	0.0301	8	2
2	molecule	0.0262	6	3
3	dopamine	0.0256	3	1
4	genetic	0.0255	5	7
5	cellular	0.0249	7	8
6	gene	0.024	4	5
7	protein	0.0239	1	4
8	neurotoxin	0.0225	26	6
9	biochemical	0.0214	22	14
10	enzyme	0.0212	20	22

Table 5.46 PI index for the quota q = 5%

№	Term	PI q = 5%	Rank	
			In-degree	BI
1	dopamine	0.0302	3	3
2	oxidation	0.0296	8	1
3	molecule	0.0287	6	2
4	protein	0.0283	1	7
5	gene	0.0276	4	6
6	neurotoxin	0.0272	25	8
7	genetic	0.0272	5	4
8	cellular	0.0265	7	5
9	synuclein	0.0247	2	18
10	serum	0.0222	46	22

Comparing the lists of drugs presented in the tables, we see that the terms "levodopa" and "neuroprotective" lead both in the general list of publications and in the Top-10 of all three centrality indices. However, six out of ten drugs no longer fall into the most cited publications, and four of them are not included in the Top-10 by all indices. Let us provide the Top-10 terms by the values of BI and PI indices for quota q = 5% (Tables 5.54 and 5.55).

Let us also provide the Top-10 terms by the values of BI and PI indices for quota q = 10% (Tables 5.56 and 5.57).

By the values of the BI and PI indices for quotas q = 5% and q = 10%, the terms "levodopa" and "neuroprotective" retain ranks 1 and 2. The Top-10 lists for the BI and PI indices with quotas q = 5% and q = 10% are practically identical.

Table 5.47 BI index for the quota q = 10%

№	Term	BI q = 10%	Rank	
			In-degree	PI
1	synuclein	0.0229	2	13
2	mpp	0.0226	36	18
3	biochemical	0.0218	21	14
4	cellular	0.0218	7	11
5	oxygen	0.0216	17	2
6	oxidation	0.0214	8	12
7	dopamine	0.0212	3	7
8	phosphorylation	0.0209	15	28
9	nucleus	0.0208	35	21
10	neurotransmitter	0.0201	34	25

Table 5.48 PI index for the quota q = 10%

№	Term	PI q = 10%	Rank	
			In-degree	BI
1	gene	0.0284	4	21
2	oxygen	0.0277	17	5
3	molecule	0.0277	6	14
4	protein	0.0266	1	12
5	sy5y	0.0242	29	27
6	enzyme	0.0238	19	30
7	dopamine	0.0235	3	7
8	plasma	0.0227	41	17
9	neurotoxin	0.022	34	10
10	genetic	0.0218	5	20

Table 5.49 Indicators for the term network medications

Name of the indicator	Values
Number of vertices	147
Number of arcs	1008
Size of maximal connected component	147
Number of connected components	1
Density	0.094

Table 5.50 Values of the In-degree index for the Top-10 terms of the Medications group

№	Terms	In-degree	Rank	
			BI	PI
1	levodopa	0.1426	2	1
2	neuroprotective	0.0858	1	2
3	amantadine	0.0538	5	4
4	rasagiline	0.0492	7	5
5	carbidopa	0.0463	9	9
6	selegiline	0.0461	12	10
7	entacapone	0.0394	19	16
8	safinamide	0.0381	27	20
9	antiparkinson	0.0365	8	6
10	pergolide	0.0343	20	18

Table 5.51 Values of the BI index for the Top-10 terms of the Medications group

№	Term	BI q = 1%	Rank	
			In-degree	PI q = 1%
1	neuroprotective	0.5344	2	2
2	levodopa	0.3304	1	1
3	inhibitory	0.045	18	7
4	haloperidol	0.0147	27	17
5	amantadine	0.0132	3	4
6	superoxide	0.0109	13	3
7	rasagiline	0.0089	4	5
8	antiparkinson	0.0087	9	6
9	carbidopa	0.0066	5	9
10	pramipexole	0.0048	11	8

5.5.2 *Term Group Symptoms*

The main indicators of the term network for the direction Symptoms are presented in Table 5.58.

The list of the Top-10 terms most frequently found in the total number of publications is presented in Table 5.59. And Table 5.60 illustrates the Top-10 terms that are found in the most cited articles. For comparison, along with the values of the In-degree index, the tables provide the ranks of the terms by BI and PI indices for quota q = 1% and k = 5.

Comparing both lists of terms (Tables 5.59 and 5.60), we see that general clinical terms—"cognitive", "dysfunction", "movement"—lead in these lists. However, the

Table 5.52 Values of the PI index for the Top-10 terms of the Medications group

№	Term	PI q = 1%	Rank	
			In-degree	BI q = 1%
1	levodopa	0.5483	1	2
2	neuroprotective	0.3635	2	1
3	superoxide	0.0179	13	6
4	amantadine	0.016	3	5
5	rasagiline	0.0114	4	7
6	antiparkinson	0.0107	9	8
7	inhibitory	0.0098	18	3
8	pramipexole	0.0045	11	10
9	carbidopa	0.0041	5	9
10	selegiline	0.0023	6	12

Table 5.53 Top-10 terms in the total number of publications and their ranks by In-degree, BI, and PI indices (q = 1%)

№	Term	Number of articles	Rank		
			In-degree	BI	PI
1	neuroprotective	2779	2	1	2
2	levodopa	2576	1	2	1
3	inhibitory	642	18	3	7
4	carbidopa	408	5	9	9
5	superoxide	369	13	6	3
6	antiparkinson	362	9	8	6
7	glutamatergic	282	17	11	14
8	insulin	264	22	13	15
9	manganese	201	31	32	30
10	immunosorbent	170	32	26	33

list of centrality terms by the In-degree index allowed us to change the composition of the list. The Top-10 included terms describing the main symptomatology of Parkinson's disease—tremor, rigidity, and bradykinesia.

Top-10 terms by the values of BI and PI indices (q = 1%) are presented in Tables 5.61 and 5.62.

The lists of terms presented in Tables 5.61 and 5.62 do not have significant differences from the list presented in Table 5.60. The exception was the term "bradykinesia", which did not enter the Top-10 by BI and PI indices, unlike the Top-10 by the In-degree index.

Table 5.54 Values of the BI index (q = 5%) for the Top-10 terms of the Medications group

№	Term	BI q = 5%	Rank	
			In-degree	PI q = 5%
1	levodopa	0.5186	1	1
2	neuroprotective	0.3007	2	2
3	inhibitory	0.0473	18	3
4	haloperidol	0.0212	27	4
5	amantadine	0.0192	3	5
6	antiparkinson	0.0165	9	7
7	rasagiline	0.0138	4	6
8	superoxide	0.0118	13	10
9	pramipexole	0.0097	11	9
10	glutamatergic	0.0051	17	8

Table 5.55 Values of the PI index (q = 5%) for the Top-10 terms of the Medications group

№	Term	PI q = 5%	Rank	
			In-degree	BI q = 5%
1	levodopa	0.5483	1	1
2	neuroprotective	0.3635	2	2
3	inhibitory	0.0179	18	3
4	haloperidol	0.016	27	4
5	amantadine	0.0114	3	5
6	rasagiline	0.0107	4	7
7	antiparkinson	0.0098	9	6
8	glutamatergic	0.0045	17	10
9	pramipexole	0.0041	11	9
10	superoxide	0.0023	13	8

Let us consider the Top-10 term lists obtained in the centrality networks by the BI and PI indices with quotas q = 5% and q = 10%. Tables 5.63 and 5.64 give the values of the BI and PI indices for quota q = 5%.

Tables 5.65 and 5.66 give the values of the BI and PI indices for quota q = 10%.

Each list presented in Tables 5.61–5.66 contains several terms that distinguish these lists from each other. For example, the term "ataxia" entered the Top-10 list for the PI index q = 10% (Table 5.66), which did not appear in the top ten by any index with quotas of 1%, 5%, and 10% except for the PI index.

Table 5.56 Values of the BI index (q = 10%) for the Top-10 terms of the Medications group

№	Term	BI q = 10%	Rank	
			In-degree	PI q = 10%
1	levodopa	0.4006	1	1
2	neuroprotective	0.3712	2	2
3	inhibitory	0.0493	18	3
4	haloperidol	0.0316	27	4
5	amantadine	0.0288	3	5
6	rasagiline	0.0230	4	7
7	antiparkinson	0.0147	9	6
8	pramipexole	0.0112	11	8
9	superoxide	0.0081	13	10
10	glutamatergic	0.0076	17	9

Table 5.57 Values of the PI index (q = 10%) for the Top-10 terms of the Medications group

№	Term	PI q = 10%	Rank	
			In-degree	BI q = 10%
1	levodopa	0.5017	1	1
2	neuroprotective	0.3325	2	2
3	inhibitory	0.0455	18	3
4	haloperidol	0.0308	27	4
5	amantadine	0.0147	3	5
6	antiparkinson	0.0136	9	7
7	rasagiline	0.0104	4	6
8	pramipexole	0.0085	11	8
9	glutamatergic	0.0055	17	10
10	superoxide	0.0041	13	9

Table 5.58 Indicators for the term network Symptoms

Name of the indicator	Value
Number of vertices	55
Number of arcs	830
Size of maximal connected component	55
Number of connected components	1
Density	0.559

Table 5.59 Total number of publications for the Top-10 terms of the Symptoms group

№	Terms	Num of publ
1	activity	7236
2	cognitive	5856
3	dysfunction	5166
4	movement	4767
5	dementia	3497
6	gait	2591
7	depressed	2389
8	sclerosis	2087
9	tremor	2083
10	atrophic	1547

Table 5.60 Values of the In-degree index for the Top-10 terms of the Symptoms group

№	Terms	In-degree	Rank q = 1%	
			BI	PI
1	movement	0.1076	1	1
2	cognitive	0.107	4	2
3	dysfunction	0.0801	2	3
4	activity	0.0707	3	14
5	tremor	0.0666	6	4
6	dementia	0.065	5	5
7	rigidity	0.0521	14	9
8	gait	0.0514	8	10
9	depressed	0.0482	11	8
10	bradykinesia	0.0448	18	15

5.5.3 Group of Terms CNS (Central Nervous System)

Network indicators of terms for the CNS direction are given in Table 5.67.

Table 5.68 presents the Top-10 CNS terms most frequently encountered in the total number of publications. For comparison, Table 5.69 presents the Top-10 terms ranked by the value of the In-degree index. This table also provides the term ranks according to the BI and PI indices (q = 1%). It is noted that the two terms "nigra" and "striatum" are leaders in both lists, i.e., they most frequently occur both in the total number of publications and in the most cited publications. The two terms "subthalamic" and "microglia", which entered the Top-10 by total number of publications (Table 5.68), dropped out of the Top-10 most cited list (Table 5.69) and were replaced by "putamen" and "thalamus".

Table 5.61 Values of the BI index for the quota q = 1% for the Top-10 terms of the Symptoms group

№	Terms	BI q = 1%	Rank	
			In-degree	PI q = 1%
1	movement	0.0764	1	1
2	dysfunction	0.0718	3	3
3	activity	0.0715	4	14
4	cognitive	0.0678	2	2
5	dementia	0.0617	6	5
6	tremor	0.0515	5	4
7	syndrome	0.0492	15	6
8	gait	0.0439	8	10
9	dyskinesia	0.0394	13	7
10	postural	0.0363	11	16

Table 5.62 Values of the PI index for the quota q = 1% for the Top-10 terms of the Symptoms group

№	Terms	PI q = 1%	Rank	
			In-degree	BI q = 1%
1	movement	0.0986	1	1
2	cognitive	0.0887	2	4
3	dysfunction	0.0887	3	2
4	tremor	0.0706	5	6
5	dementia	0.0605	6	5
6	syndrome	0.0586	15	7
7	dyskinesia	0.0496	13	9
8	depressed	0.0458	9	11
9	rigidity	0.0451	7	14
10	gait	0.036	8	8

In the following Tables 5.70 and 5.71, a list of terms included in the Top-10 by BI and PI indices (q = 1%) is given.

Attention should be drawn to the differences of these lists from those presented in Table 5.68. The two terms "nigra" and "striatum" lost their leading positions, and the term "striatum" dropped out of the Top-10 by the PI index (Table 5.71). The leader in the Top-10 terms by the influence index on pivotal vertices PI became the term "subcortical", which had not previously entered the Top-10 by frequency of mention in the total number of works and in highly cited publications. Tables 5.72 and 5.73 present the BI and PI index values (quota q = 5%), and Tables 5.74 and 5.75 (quota q = 10%) for the Top-10 terms of the CNS group.

Table 5.63 Values of the BI index for the quota q = 5% for the Top-10 terms of the Symptoms group

№	Terms	BI q = 5%	Rank	
			In-degree	PI q = 5%
1	movement	0.0773	1	1
2	dysfunction	0.0682	3	2
3	activity	0.0675	4	24
4	cognitive	0.065	2	3
5	syndrome	0.059	15	6
6	tremor	0.0562	5	4
7	dementia	0.0551	6	14
8	dyskinesia	0.0459	13	7
9	gait	0.0429	8	10
10	depressed	0.0363	9	9

Table 5.64 Values of the PI index for the quota q = 5% for the Top-10 terms of the Symptoms group

№	Terms	PI q = 5%	Rank	
			In-degree	BI q = 5%
1	movement	0.0866	1	1
2	dysfunction	0.0787	3	2
3	cognitive	0.0746	2	4
4	tremor	0.0668	5	6
5	dystonia	0.0612	22	14
6	syndrome	0.0576	15	5
7	dyskinesia	0.044	13	8
8	rigidity	0.0426	7	11
9	depressed	0.0407	9	10
10	gait	0.032	8	9

As can be seen from the presented tables, for this group of terms as well, differences in the Top-10 lists for various indices and quotas are visible.

5.5.4 *Group of Terms Diagnosis*

Table 5.76 presents the network indicators of terms for the Diagnosis direction.

Table 5.65 Values of the BI index for the quota q = 10% for the Top-10 terms of the Symptoms group

№	Terms	BI q = 10%	Rank	
			In-degree	PI q = 10%
1	dysfunction	0.0703	3	2
2	movement	0.0701	1	1
3	cognitive	0.0651	2	4
4	activity	0.0624	4	23
5	syndrome	0.0579	15	3
6	dementia	0.0566	6	15
7	tremor	0.0536	5	5
8	gait	0.0411	8	12
9	dyskinesia	0.0392	13	6
10	postural	0.0384	11	18

Table 5.66 Values of the PI index for the quota q = 10% for the Top-10 terms of the Symptoms group

№	Terms	PI q = 10%	Rank	
			In-degree	BI q = 10%
1	movement	0.0778	1	2
2	dysfunction	0.0702	3	1
3	syndrome	0.0692	15	5
4	cognitive	0.0613	2	3
5	tremor	0.0603	5	7
6	dyskinesia	0.0524	13	9
7	dystonia	0.0508	22	15
8	depressed	0.0383	9	12
9	ataxia	0.038	25	19
10	rigidity	0.0377	7	11

Table 5.67 Indicators for the term network CNS

Name of the indicator	Value
Number of vertices	39
Number of arcs	568
Size of maximal connected component	39
Number of connected components	1
Density	0.767

Table 5.68 Total number of publications for the Top-10 terms of the CNS group

№	Terms	Num of publ
1	nigra	4290
2	striatum	3527
3	mitochondria	3089
4	cortex	2984
5	basal	1750
6	subthalamic	1700
7	ganglia	1584
8	microglia	1553
9	nigrostriatal	1326
10	midbrain	1287

Table 5.69 Indices for Top-10 terms of the CNS group

№	Terms	In-degree	Rank, q = 1%	
			BI	PI
1	nigra	0.1268	5	9
2	striatum	0.1101	7	11
3	cortex	0.0993	3	5
4	basal	0.0763	1	3
5	ganglia	0.0714	2	2
6	nigrostriatal	0.0479	6	6
7	mitochondria	0.0436	16	20
8	midbrain	0.0418	11	16
9	putamen	0.0402	12	7
10	thalamus	0.0393	4	18

Table 5.70 BI index values for the quota q = 1% for the Top-10 terms of the CNS group

№	Terms	BI q = 1%	Rank	
			In-degree	PI q = 1%
1	basal	0.0641	4	3
2	ganglia	0.0619	5	2
3	cortex	0.0571	3	5
4	thalamus	0.0571	10	18
5	nigra	0.057	1	9
6	nigrostriatal	0.0564	6	6
7	striatum	0.05	2	11
8	cerebral	0.0492	13	8
9	subcortical	0.0427	14	1
10	cerebellar	0.0424	17	28

Table 5.71 PI index values for the quota q = 1% for the Top-10 terms of the CNS group

№	Terms	PI q = 1%	Rank	
			In-degree	BI q = 1%
1	subcortical	0.0785	14	9
2	ganglia	0.0721	5	2
3	basal	0.0704	4	1
4	hippocampus	0.0679	15	13
5	cortex	0.0599	3	3
6	nigrostriatal	0.057	6	6
7	putamen	0.0541	9	12
8	cerebral	0.0529	13	8
9	nigra	0.0517	1	5
10	neurogenesis	0.0477	26	19

Table 5.72 BI index values for the quota q = 5% for the Top-10 terms of the CNS group

№	Terms	BI q = 5%	Rank	
			In-degree	PI q = 5%
1	cortex	0.0654	3	4
2	thalamus	0.0641	10	2
3	ganglia	0.0608	5	3
4	nigrostriatal	0.058	6	5
5	basal	0.0561	4	1
6	striatum	0.0549	2	6
7	nigra	0.0525	1	8
8	cerebellar	0.0481	17	9
9	cerebral	0.0471	13	7
10	subcortical	0.0416	14	16

Let us recall that since publications dedicated to Parkinson's disease were analyzed, the diagnostic term "parkinson" was excluded from the analysis. In the list of Top-10 terms related to disease names, the term "alzheimer" was most frequently encountered in the publications. The full list of Top-10 terms is given in Table 5.77. For comparison, Table 5.78 presents the list of Top-10 terms that occur in the most cited publications.

Table 5.73 PI index values for the quota q = 5% for the Top-10 terms of the CNS group

№	Terms	PI q = 5%	Rank	
			In-degree	BI q = 5%
1	basal	0.0802	4	5
2	thalamus	0.077	10	2
3	ganglia	0.0765	5	3
4	cortex	0.0682	3	1
5	nigrostriatal	0.066	6	4
6	striatum	0.0628	2	6
7	cerebral	0.0537	13	9
8	nigra	0.0528	1	7
9	cerebellar	0.0503	17	8
10	supranuclear	0.0405	19	14

Table 5.74 BI index values for the quota q = 10% for the Top-10 terms of the CNS group

№	Terms	BI q = 10%	Rank	
			In-degree	PI q = 10%
1	thalamus	0.0667	10	1
2	cortex	0.0645	3	3
3	nigra	0.0571	1	9
4	basal	0.0569	4	5
5	striatum	0.0549	2	6
6	nigrostriatal	0.0537	6	7
7	ganglia	0.0521	5	4
8	cerebellar	0.0521	17	2
9	cerebral	0.0503	13	8
10	supranuclear	0.042	19	12

As can be seen from Tables 5.77 and 5.78, Alzheimer's disease (AD) diagnosis ranks first. This allows us to conclude that in publications dedicated to Parkinson's disease, AD holds a significant place. The second most mentioned disease in publications is Lewy body disease ("lewy"). Since 23 terms were analyzed in this section, the Top-10 terms we consider represent nearly half of the term set. It might be assumed that networks built on three centrality indices would not differ significantly. However, the comparative analysis of the terms presented in the tables shows that in this case as well, centrality indices allow us to obtain some new information. Tables 5.79 and

Table 5.75 PI index values for the quota q = 10% for the Top-10 terms of the CNS group

№	Terms	PI q = 10%	Rank	
			In-degree	BI q = 10%
1	thalamus	0.0805	10	1
2	cerebellar	0.0788	17	8
3	cortex	0.0659	3	2
4	ganglia	0.0633	5	7
5	basal	0.0632	4	4
6	striatum	0.0624	2	5
7	nigrostriatal	0.0613	6	6
8	cerebral	0.0572	13	9
9	nigra	0.0478	1	3
10	pyramidal	0.0419	34	19

Table 5.76 Indicators for the term network Diagnosis

Name of the indicator	Values
Number of vertices	23
Number of arcs	172
Size of maximal connected component	23
Number of connected components	1
Density	0.680

Table 5.77 Total number of publications for the Top-10 terms of the Diagnosis group

№	Terms	Num of publ
1	alzheimer	5539
2	lewy	2486
3	huntington	1382
4	amyotrophic	1371
5	cancer	747
6	palsy	663
7	msa	656
8	schizophrenia	613
9	supranuclear	612
10	prion	535

Table 5.78 In-degree index values for the Top-10 terms in the Diagnosis group

№	Terms	In-degree	Rank, q = 1%	
			BI	PI
1	alzheimer	0.1939	1	1
2	lewy	0.1673	2	3
3	supranuclear	0.0821	4	5
4	palsy	0.0802	5	6
5	msa	0.0773	8	4
6	amyotrophic	0.0752	3	2
7	huntington	0.0702	6	10
8	synucleinopathy	0.048	9	7
9	prion	0.0429	15	9
10	pdd	0.0378	13	12

5.80 present the BI and PI index values (q = 1%) for the Top-10 terms of the Diagnosis group. As seen from Tables 5.79 and 5.80, the term “schizophrenia”, included in the Top-10 by mentions in the total number of publications (Table 5.77), does not enter the Top-10 by any centrality index. Conversely, the term “synucleinopathy” is included in the Top-10 of all three indices despite being absent in the Top-10 of total number of publications. Thus, the term “cancer”, ranked in the top five by number of mentions in the total number of publications, appears in the tenth position in the network based on the group influence index BI, q = 1% (Table 5.79).

Table 5.79 BI index values for the quota q = 1% for the Top-10 terms of the Diagnosis group

№	Terms	BI q = 1%	Rank	
			In-degree	PI q = 1%
1	alzheimer	0.1637	1	1
2	lewy	0.0995	2	3
3	amyotrophic	0.0826	6	2
4	supranuclear	0.0759	3	5
5	palsy	0.0759	4	6
6	huntington	0.0747	7	10
7	ipd	0.0738	18	11
8	msa	0.0737	5	4
9	synucleinopathy	0.0525	8	7
10	cancer	0.0435	16	17

Table 5.80 PI index values for the quota q = 1% for the Top-10 terms of the Diagnosis group

№	Terms	PI q = 1%	Rank	
			In-degree	BI q = 1%
1	alzheimer	0.1855	1	1
2	amyotrophic	0.1648	6	3
3	lewy	0.127	2	2
4	msa	0.111	5	8
5	supranuclear	0.1093	3	4
6	palsy	0.1067	4	5
7	synucleinopathy	0.0853	8	9
8	tauopathy	0.0451	12	12
9	prion	0.0157	9	15
10	huntington	0.0151	7	6

Table 5.81 BI index values for the quota q = 5% for the Top 10 terms of the Diagnosis group

№	Terms	BI q = 5%	Rank	
			In-degree	PI q = 5%
1	alzheimer	0.1635	1	1
2	lewy	0.1018	2	3
3	huntington	0.0852	7	2
4	palsy	0.0788	4	5
5	supranuclear	0.0788	3	6
6	ipd	0.0763	18	7
7	msa	0.0704	5	4
8	amyotrophic	0.0566	6	9
9	synucleinopathy	0.0535	8	8
10	cbd	0.0466	13	13

The Top 10 terms by BI and PI indices (quota q = 5%) are presented in Tables 5.81 and 5.82.

Tables 5.83 and 5.84 give the Top-10 BI and PI terms with a 10% quota.

5.5.5 *Group of Terms Research Methods*

Table 5.85 presents the indicators of the term network for the direction Research Methods.

Table 5.82 PI index values for the quota q = 5% for the Top 10 terms of the Diagnosis group

№	Terms	PI q = 5%	Rank	
			In-degree	BI q = 5%
1	alzheimer	0.1384	1	1
2	huntington	0.1376	7	3
3	lewy	0.0897	2	2
4	msa	0.0829	5	7
5	palsy	0.0796	4	4
6	supranuclear	0.0796	3	5
7	ipd	0.0677	18	6
8	synucleinopathy	0.0637	8	9
9	amyotrophic	0.056	6	8
10	pdd	0.0417	10	12

Table 5.83 BI index values for the quota q = 10% for the Top-10 terms of the Diagnosis group

№	Terms	BI q = 10%	Rank	
			In-degree	PI q = 10%
1	alzheimer	0.1447	1	1
2	lewy	0.1033	2	6
3	huntington	0.0871	7	5
4	supranuclear	0.085	3	4
5	palsy	0.0834	4	3
6	ipd	0.0705	18	2
7	msa	0.0677	5	7
8	amyotrophic	0.0594	6	10
9	cbd	0.0573	13	11
10	synucleinopathy	0.0422	8	9

This network has 20 nodes. Half of the terms in this network are presented in the Top-10 tables for three indices.

Table 5.87 provides the values of the In-degree index, Table 5.88—the values for BI (q = 1%), and the values for PI (q = 1%) for the Top-10 terms of this network are shown in Table 5.89. The tables contain not only the values of the indices themselves but also the rankings of the other considered indices. For comparison, Table 5.86 presents the Top-10 terms by the total number of publications. All terms included in this table are ranked in Tables 5.87–5.89. However, it should be noted that new

Table 5.84 PI index values for the quota q = 10% for the Top-10 terms of the Diagnosis group

№	Terms	PI q = 10%	Rank	
			In-degree	BI q = 10%
1	alzheimer	0.1267	1	1
2	ipd	0.0946	18	6
3	palsy	0.0935	4	5
4	supranuclear	0.0904	3	4
5	huntington	0.084	7	3
6	lewy	0.0821	2	2
7	msa	0.0758	5	7
8	cancer	0.0592	16	11
9	synucleinopathy	0.0583	8	10
10	amyotrophic	0.0513	6	8

Table 5.85 Indicators for the term network Research Methods

Name of the indicator	Values
Number of vertices	20
Number of arcs	136
Size of maximal connected component	20
Number of connected components	1
Density	0.716

terms also appear in the tables with network terms. According to the BI index, this includes "electroencephalogram", and for the PI index—"123i". Both terms did not enter the Top-10 by the total number of publications (Table 5.86) The term "electroencephalogram" is mentioned in 203 works, and the term "123i"—in 241 publications. It can be assumed that the term "electroencephalogram", which entered the Top-10 by the group influence index BI, is frequently encountered in highly cited publications in narrowly specialized groups.

Tables 5.88 and 5.89 present the index values for the Top-10 Terms of the Group "Research Methods".

The BI and PI index values for the quotas q = 5% for the Top-10 terms are presented in Tables 5.90–5.91, and for the quotas q = 10% in Tables 5.92–5.93.

Table 5.86 Total number of articles for the Top-10 terms of the group Research Methods

№	Terms	Number of articles
1	stimulation	3661
2	updrs	2222
3	dbs	2009
4	mri	1368
5	surgery	1017
6	tomography	856
7	neuroimaging	664
8	pdq	413
9	transplantation	405
10	fmri	374

Table 5.87 In-degree index for the Top-10 terms of the group Research Methods

№	Terms	In-degree	Rank, q = 1%	
			BI	PI
1	stimulation	0.2501	3	2
2	dbs	0.2075	5	3
3	updrs	0.1206	2	1
4	surgery	0.1014	7	5
5	mri	0.0743	1	12
6	tomography	0.0453	6	6
7	neuroimaging	0.0417	4	9
8	pdq	0.0342	9	8
9	fmri	0.0189	14	11
10	mmse	0.0189	8	13

Table 5.88 BI index for the quota q = 1% for the Top-10 terms of the group Research Methods

№	Terms	BI q = 1%	Rank	
			In-degree	PI q = 1%
1	mri	0.1375	5	12
2	updrs	0.1341	3	1
3	stimulation	0.1325	1	2
4	neuroimaging	0.1015	7	9
5	dbs	0.0866	2	3
6	tomography	0.0714	6	6
7	surgery	0.063	4	5
8	mmse	0.0479	10	4
9	pdq	0.0373	8	8
10	electroencephalogram	0.0343	14	15

Table 5.89 PI index for the quota q = 1% for the Top-10 terms of the group Research Methods

№	Terms	PI q = 1%	Rank	
			In-degree	BI q = 1%
1	updrs	0.2506	3	2
2	stimulation	0.2066	1	3
3	dbs	0.1561	2	5
4	mmse	0.1328	10	8
5	surgery	0.1041	4	7
6	tomography	0.0692	6	6
7	transplantation	0.0291	13	15
8	pdq	0.012	8	9
9	neuroimaging	0.009	7	4
10	123i	0.0081	11	13

Table 5.90 BI index for the quota q = 5% for the Top-10 terms of the group Research Methods

№	Terms	BI q = 5%	Rank	
			In-degree	PI q = 5%
1	mri	0.1551	6	5
2	updrs	0.1431	3	1
3	stimulation	0.1159	1	3
4	neuroimaging	0.1129	7	6
5	tomography	0.0759	6	7
6	dbs	0.0751	2	2
7	surgery	0.0503	4	4
8	mmse	0.0497	10	8
9	pdq	0.039	8	9
10	electroencephalogram	0.031	14	12

Table 5.91 PI index for the quota q = 5% for the Top-10 terms of the group Research Methods

№	Terms	PI q = 5%	Rank	
			In-degree	BI q = 5%
1	updrs	0.1937	3	2
2	dbs	0.1457	2	6
3	stimulation	0.1082	1	3
4	surgery	0.0972	4	7
5	mri	0.0774	5	1
6	neuroimaging	0.0658	7	4
7	tomography	0.0646	6	5
8	mmse	0.0486	10	8
9	pdq	0.0443	8	9
10	fmri	0.0319	9	14

Table 5.92 BI index for the quota q = 10% for the Top-10 terms of the group Research Methods

№	Terms	BI q = 10%	Rank	
			In-degree	PI q = 10%
1	mri	0.1584	5	2
2	updrs	0.1437	3	1
3	stimulation	0.1271	1	5
4	neuroimaging	0.1141	7	3
5	dbs	0.0824	2	4
6	tomography	0.0661	6	7
7	surgery	0.0474	4	6
8	mmse	0.0432	10	10
9	electro-encephalogram	0.034	14	9
10	pdq	0.0324	8	8

Table 5.93 PI index for the quota q = 10% for the Top-10 terms of the group Research Methods

№	Terms	PI q = 10%	Rank	
			In-degree	BI q = 10%
1	updrs	0.1482	3	2
2	mri	0.1425	5	1
3	neuroimaging	0.1244	7	4
4	dbs	0.1115	2	5
5	stimulation	0.0828	1	3
6	surgery	0.0743	4	7
7	tomography	0.0494	6	6
8	pdq	0.0458	8	10
9	electro-encephalogram	0.0377	14	9
10	mmse	0.0372	10	8

References

Ashoka, R., & Harinarayana, N. S. (2022). Keyword analysis of research trends in 'scientometrics' journal from 2014 to 2018. *Journal of Information and Knowledge, 59*(3), 163–169. https://doi.org/10.17821/srels/2022/v59i3/165826

Cinpolat, H. Y., & Akar, A. (2023). A visual and bibliometric analysis of Parkinson's disease biomarker research. *Annals of Medical Research, 30*(10), 1359–1364. https://doi.org/10.5455/annalsmedres.2023.09.253

Fan, F. (2024). The new science of COVID-19: A bibliographic and network analysis. arXiv:2407.15867 [cs.DL]. https://doi.org/10.48550/arXiv.2407.15867

Liao, X. Y., Gao, Y. X., Qian, T. T., Zhou, L.-H., Li, L.-Q., Gong, Y., & Ye, T. F. (2024). Bibliometric analysis of electroencephalogram research in Parkinson's disease from 2004 to 2023. *Frontiers in Neuroscience: Neurodegeneration, 18*, 1433583. https://doi.org/10.3389/fnins.2024.1433583

Lin, T. Z., Tang, K. U., Lin, S. S., Changlai, M.-L., & Hsu, Y.-S. (2022). Co-word analysis of selected science education literature: Identifying research trends of scaffolding in two decades (2000–2019). *Frontiers in Psychology, 13*, Article 844425. https://doi.org/10.3389/fpsyg.2022.844425

Lozano, S., Calzada-Infante, L., Adenso-Díaz, B., & García, S. (2019). Complex network analysis of keywords co-occurrence in the recent efficiency analysis literature. *Scientometrics, 120*, 609–629. https://doi.org/10.1007/s11192-019-03132-w

Molontay, R., & Nagy, M. (2020). Twenty years of network science: A bibliographic and co-authorship network analysis. arXiv:2001.09006 [physics.soc-ph], https://doi.org/10.48550/arXiv.2001.09006

Shi, L., & Yih, B. (2024). Knowledge mapping and research trends of accidental falls in patients with Parkinson's disease from 2003 to 2023: A bibliometric analysis. *Frontiers in Neurology, 15*, 1443799. https://doi.org/10.3389/fneur.2024.1443799

Utegenova, A. B., Yermagambetova, A. P., Kabdrakhmanova, G. B., Khamidulla, A. A., Urasheva, Z. U., Kenzhina, N. K., Zhussupova, Z., & Nurlanova, G. N. (2025). Bibliometric analysis of alpha-synuclein determination by biopsy in peripheral tissues of patients with Parkinson's disease. *International Journal of Telemedicine and Applications, 2025*, 6469893. https://doi.org/10.1155/ijta/6469893

Yazdanjue, N., Yazdanjouei, H., Gharoun, H., Khorshidi, M. S., Rakhshaninejad, M., Amiri, B., & Gandomi, A. H. (2025). A comprehensive bibliometric analysis on social network anonymization: Current approaches and future directions. *Knowledge and Information Systems, 67*, 29–108. https://doi.org/10.1007/s10115-024-02289-y

Zhang, T., Yang, R., Pan, J., & Huang, S. (2023). Parkinson's disease related depression and anxiety: A 22-year bibliometric analysis (2000–2022). *Neuropsychiatric Disease and Treatment, 19*, 1477–1489. https://doi.org/10.2147/NDT.S403002

Zhu, L., Zhu, P., Wang, J., Yan, K., Zhao, S., Jiang, Y., & Zhang, H. (2024). A bibliometric and visual analysis of Parkinson's disease sleep disorders: Articles from 2008 to 2023. *Frontiers in Psychiatry, 15*, 1468568. https://doi.org/10.3389/fpsyt.2024.1468568

Chapter 6
An Analysis of Journals Citations

Particular attention in bibliometric analysis is given to identifying the most influential journals, since they serve as the main channels for the dissemination of scientific knowledge. The phenomenon of citation is undeniably an important principle in science, and, undoubtedly, one of the important means of scientific communication. Citation becomes a standard ethical norm in science approximately in the middle of the last century, when the scientific journal begins to be considered a social institution and an integral tool of communication among scientists. Citation indicators include the idea of evaluating two aspects of a journal: its productivity and scientific popularity. Determining the rank of a journal in a given research field helps in assessing the prospects of publishing in that journal.

In (Merigó & Yang, 2017), a bibliometric review of journals in the field of accounting based on Web of Science data is presented. The authors identified the most influential journals with a significant share of citations in the field of accounting. The study identified major trends in scientific publications, including the growth in the number of studies, the use of various methodological approaches, and the impact of international collaboration on the development of the discipline. The study emphasizes the importance of bibliometric analysis for assessing the structure and dynamics of scientific research in the field of accounting.

In (Ioannidis, 2006), the concentration of the most cited works in various scientific fields is considered, and it is concluded that only 9% of all journals publish the majority of the most influential articles. Moreover, a significant portion of the most cited publications is published in a small number of multidisciplinary journals, such as "Nature" and "The Lancet." This confirms the hypothesis of the uneven distribution of the influence of scientific journals. The article uses statistical analysis to assess the concentration of the most cited works in different scientific fields.

Classical metrics for assessing the influence of a journal, such as the Journal Impact Factor (JIF), remain widely used but are increasingly criticized. For example, (Stachura et al., 2023) analyzed discrepancies between JIF and expert evaluations of journals in Poland. The study showed that in a number of cases, journals with

© The Editor(s) (if applicable) and The Author(s), under exclusive license to Springer Nature Switzerland AG 2026

F. Aleskerov et al., *Bibliometric Analysis by Network Models*, Contemporary Systems Thinking, https://doi.org/10.1007/978-3-032-09171-0_6

low JIF received high expert evaluations, which states the question about the objectivity of traditional bibliometric indicators in evaluating scientific achievements. The COVID-19 pandemic also had a significant impact on the bibliometric indicators of leading medical journals. An analysis of six popular and influential medical journals showed that the number of citations of publications related to COVID-19 significantly exceeded the indicators of other topics (Delardas & Giannos, 2021). This led to a sharp increase in the impact factor of these journals, which cast doubt on its reliability as a long-term indicator of scientific influence. Traditional metrics such as JIF allow the identification of key journals, but their reliability is questioned due to the influence of global events and subjective factors. Alternative methods, such as three-dimensional models for journal assessment and more complex network methods of citation analysis, can provide a more objective picture of scientific influence. Thus, in (Zeng et al., 2024), a three-dimensional approach to journal assessment is proposed, which takes into account article citation, author influence, and institutional influence. The use of the CRITIC-Entropy method for weighted analysis showed that such an approach yields more accurate and objective results compared to traditional indicators such as JIF. This method allows for a more balanced assessment of the influence of scientific journals, especially in the economic field.

In (Cowhitt & Cutts, 2020), the authors used network analysis to assess the degree of journal coverage by various bibliographic databases in the field of emergency medicine. The study used bibliometric data to build networks representing the interconnections of journals based on citation patterns. In particular, the authors applied bibliographic coupling and co-citation analysis to assess the similarity and interconnectedness between journals. These methods made it possible to identify the main journals and visualize their positions in the network. Metrics such as degree centrality and betweenness centrality were used to quantitatively assess the influence and bridging roles of individual journals. Specialized software tools were used to analyze the data, including VOSviewer and Gephi, which are designed for constructiong and visualizing bibliometric networks.

In this chapter, we consider the evaluation of journal publication activity using methods of statistical and network analysis.

6.1 Statistical Analysis of the Publication Activity of Journals

From the downloaded list of publications, as described above, 39,811 articles on the study of various aspects of Parkinson's disease (PD) published between 2015 and 2021 in 3,292 journals were retained for analysis through filtering. During the entire period under review, 567 journals received no citations. The distribution of citations across the remaining journals is shown in the graph in Fig. 6.1.

As shown in Fig. 6.1, most journals (1,426) had between 1 and 9 citations. Twenty-four journals had 1,000 or more citations.

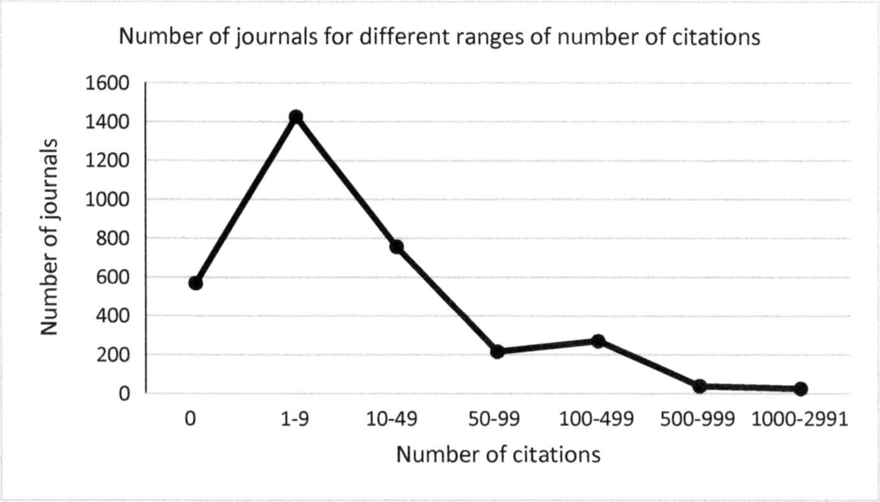

Fig. 6.1 Number of citations to journals in the period 2015–2021

The analysis of the publication activity of journals revealed that 1,260 journals published one article each during the entire period. Notably, the interest in PD publications among these multidisciplinary journals varied over the years (Fig. 6.2).

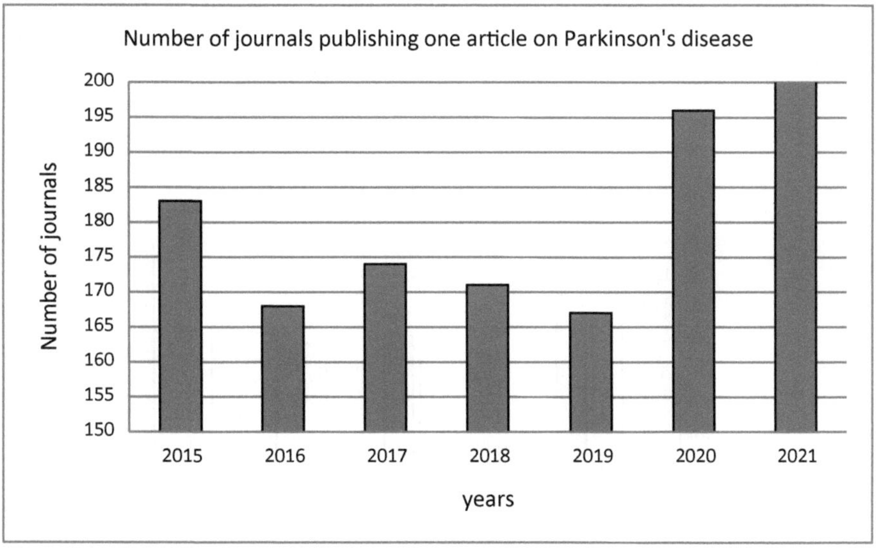

Fig. 6.2 Diagram showing the number of journals publishing one article per year on PD

In the same period (2015–2021), 68 journals published between 100 and 1,000 articles, and only 2 journals published more than 1,000 articles. These two academic journals are:

- "Parkinsonism & Related Disorders" (ISSN: 18735126, 13538020; published by Elsevier BV)
- "Movement Disorders" (ISSN: 08853185, 15318257), which is the official journal of the International Parkinson and Movement Disorder Society (MDS).

Both journals are widely read and cover all topics in the field—both clinical and fundamental sciences.

The journal Parkinsonism & Related Disorders, which publishes results of fundamental and clinical research on all neurodegenerative syndromes, including Parkinson's disease (PD), had 1,260 publications on PD over the entire period. The journal's Quartile is Q1, h-index is 113 (Scopus), and its Impact Factor is 3.184 (Scopus), 4.891 (WoS). The highest SJR score was 1.753 in 2017, while the lowest was 1.093 in 2021.

The journal Movement Disorders mainly publishes articles on Parkinson's disease and dystonia. Among the publications we reviewed, 1,201 articles were published in this journal. The journal's main bibliometric indicators are: Quartile Q1, h-index 219 (Scopus), Impact Factor 5.67 (Scopus). The lowest Impact Score for the journal was 4.41 in 2015, while the highest was 5.90 in 2017. The SJR was highest in 2018 at 3.622 and lowest in 2021 at 2.473.

For the two leading journals by the number of PD publications in the period 2015–2021, the distribution of publications by year is shown in Fig. 6.3.

As seen in Fig. 6.3, the number of PD publications in these journals has fluctuated slightly from year to year without any sharply defined trends in either direction.

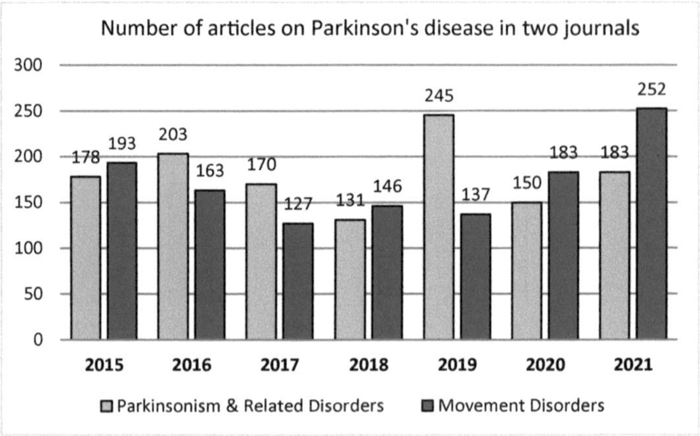

Fig. 6.3 Number of PD publications in each year in Parkinsonism & Related Disorders and Movement Disorders

In third place in publication activity was bioRxiv, a preprint server. We downloaded 795 PD-related publications from the server. The server is not indexed and, furthermore, does not have an impact factor.

Following in the number of publications next journals are

- Journal of Parkinson's Disease (JPD) (ISSN: 1877718X, 18777171), published by IOS Press in Amsterdam (The Netherlands)—768 publications (Q1, h-index 60 (Scopus), SJR 1.325), and
- Scientific Reports (ISSN 20452322, published by Nature Publishing Group, UK)—696 publications (Q1, h-index 282 (Scopus), Impact Factor 4.44 (Scopus)).

These are the Top-5 journals in terms of the number of publications that we downloaded from the database.

During this period (2015–2021), some journals reduced their publication activity, while others increased the number of publications.

For example, the international interdisciplinary peer-reviewed open-access scientific journal "PLOS One" (ISSN 19326203), which ranks seventh in the number of publications on Parkinson's disease (670 articles over seven years), with an h-index of 404 (Scopus), an Impact Factor of 3.75 (Scopus), and Q1, consistently decreased the number of publications on Parkinson's disease.

Fig. 6.4 shows a diagram illustrating the decline in the number of publications in the journal.

An example of a journal that increased its number of publications is "Frontiers in Aging Neuroscience" published by Frontiers Media S.A., Switzerland (ISSN 16634365, ranked 13th in our database by the number of publications—678 articles on Parkinson's disease). The journal had an h-index of 105 (Scopus), Q1, and

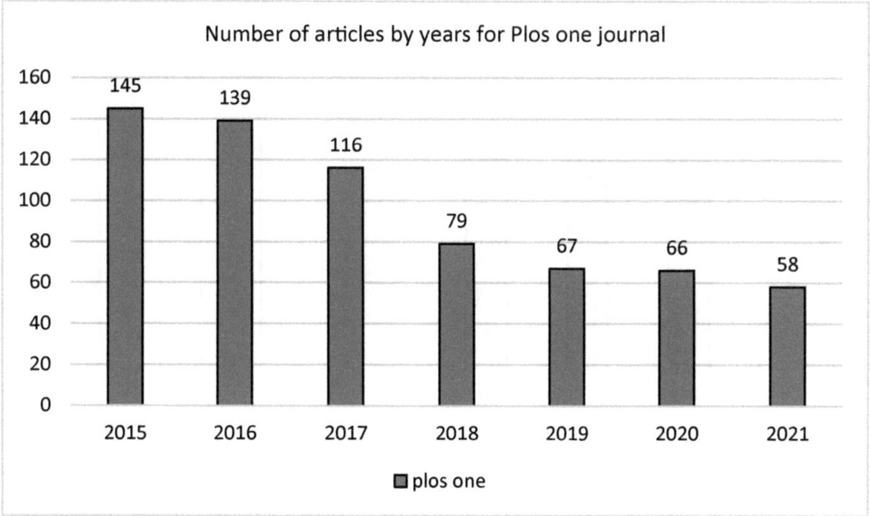

Fig. 6.4 Number of publications in the journal Plos One on Parkinson's disease each year

SJR 1.324 (2021). Fig. 6.5 shows a diagram of the number of publications in the journal each year under review.

It is worth noting that the total number of publications on Parkinson's disease grew from 2015 to 2021. Fig. 6.6 shows a graph reflecting the growth in the total number of publications.

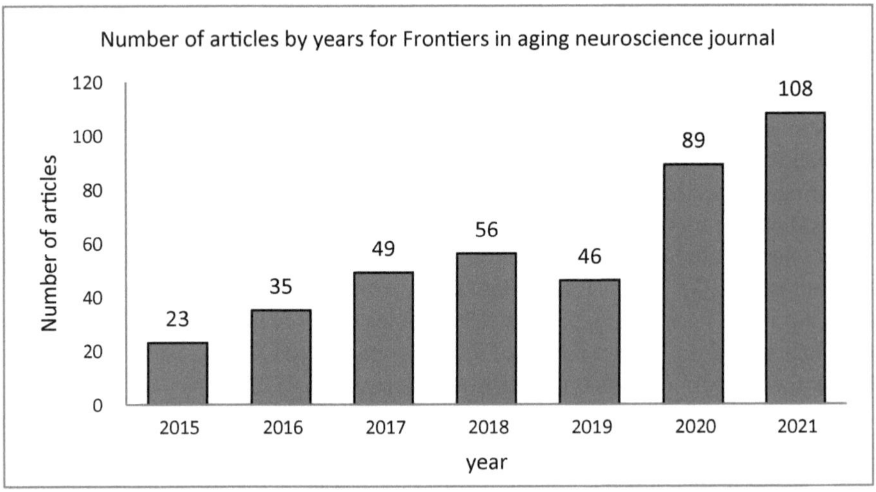

Fig. 6.5 Number of publications in the journal "Frontiers in Aging Neuroscience" on Parkinson's disease each year

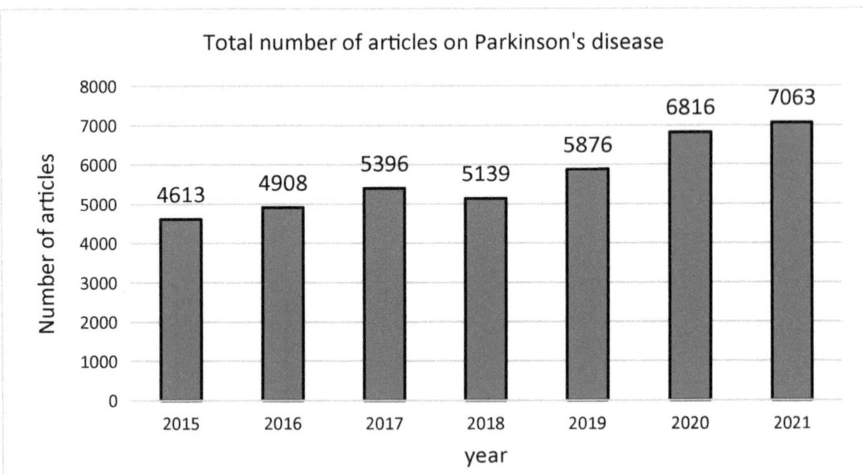

Fig. 6.6 Dynamics of the growth in the number of publications on Parkinson's disease during the reviewed period

Next, let us consider and analyze the citation network of journals, based on the citation of articles published in these journals, or the citation network of journals.

6.2 Journal Citation Network

Various citation analysis methods can be reduced to two classes: (1) citation statistics and (2) citation network analysis. Citation metrics include the assessment of two aspects of a journal's activity—its productivity and scientific popularity.

It is important to emphasize that quantitative data on the citation of publications reflect the impact of the research outcome on the scientific community, its usefulness to other scientists. These data alone do not measure the quality of the publication; therefore, they should be considered as indicators showing that the work may be highly significant with a certain probability. Citation counting extends the possibilities of bibliometric evaluations alongside and in conjunction with other indicators.

Citation analysis of journals allows identifying the most influential publications and highlighting potential citation clusters formed among journals. Information about each publication includes the journal (publication) in which it was published, allowing the construction of a network of interactions between scientific journals on a particular topic. The graph construction proceeds as follows: the vertex represents a journal, and an edge between vertices indicates that an article published in journal A cites an article in journal B.

For describing the constructed network, the following attributes are used

- JId—Journal ID;
- Y—Year of journal publication;
- Weight—Weight of the edge.

The network represents a weighted graph consisting of 3292 journals and 152203 edges. An edge represents the citation of an article from journal JId1 to an article in journal JId2. The weight of the edge is the number of citations from journal JId1 to journal JId2 for a particular year (Fig. 6.7).

JId1	JId2	Weight	Year
10623703	10623703	101	2021
10623703	163027424	150	2021
115201623	163027424	143	2020
115201623	163027424	134	2021
118428158	163027424	146	2020
118428158	163027424	127	2021
147691530	147691530	120	2017

Fig. 6.7 A fragment of the journal citation network. The columns JId1 and JId2 represent the journal IDs listed in the database

Table 6.1 Metrics for the journal citation network

Name of the indicator	Value
Number of nodes	3292
Number of edges	152203
Size of the largest connected component	3285
Number of connected components	4
Density	0.0094
Maximum number of citations	24122
Minimum number of citations	1
Average number of citations	113.5

The Weight column represents the number of citations in journal JId1 to journal JId2; the "Year" column indicates the year in which citations are reflected. The main metrics of the journals citation network are presented in Table 6.1.

The largest connected component of the four contained 3285 journals. The second component contained 3 journals, and the third and fourth components each contained 2 journals. The structure of the journal citation graph differs from other types of considered graphs due to the presence of self-citations. A publication cannot have self-citations, but at the same time, an article in journal A may cite other publications in the same journal A. Self-citation in journals is a common phenomenon; for example, in the journal "Brain" (ISSN: 14602156, 00068950; ID—118357697), 18% (248) of citations during the review period are to publications from the same journal.

We conducted a separate analysis of journal citation rates.

The number of references to the journal that published the cited article was considered, as well as to the journal in which the citing article was published. The number of citing articles for each publication in the given journal was counted. Self-citation of journals was also analyzed. The analysis was carried out for the entire period as well as for each year under consideration.

Fig. 6.8 shows the Top-11 journals with the highest number of references to them over the entire period under consideration (2015–2021), excluding self-citation. The diagram illustrates the number of articles and journals in which these articles were published, citing publications in the given journal.

We will now consider statistics indicating the number of journals citing other journals in their publications. The journal-to-journal citation analysis allowed us to identify 24 journals that were cited by at least 1000 other journals over 7 years. Table 6.2 lists these journals. Various publications in 427 journals received one citation in other journals' publications during this period.

It is worth noting that both journals with the largest number of publications—"Movement Disorders" (1201 articles) and "Parkinsonism & Related Disorders" (1260 articles)—also rank first in terms of the number of citations.

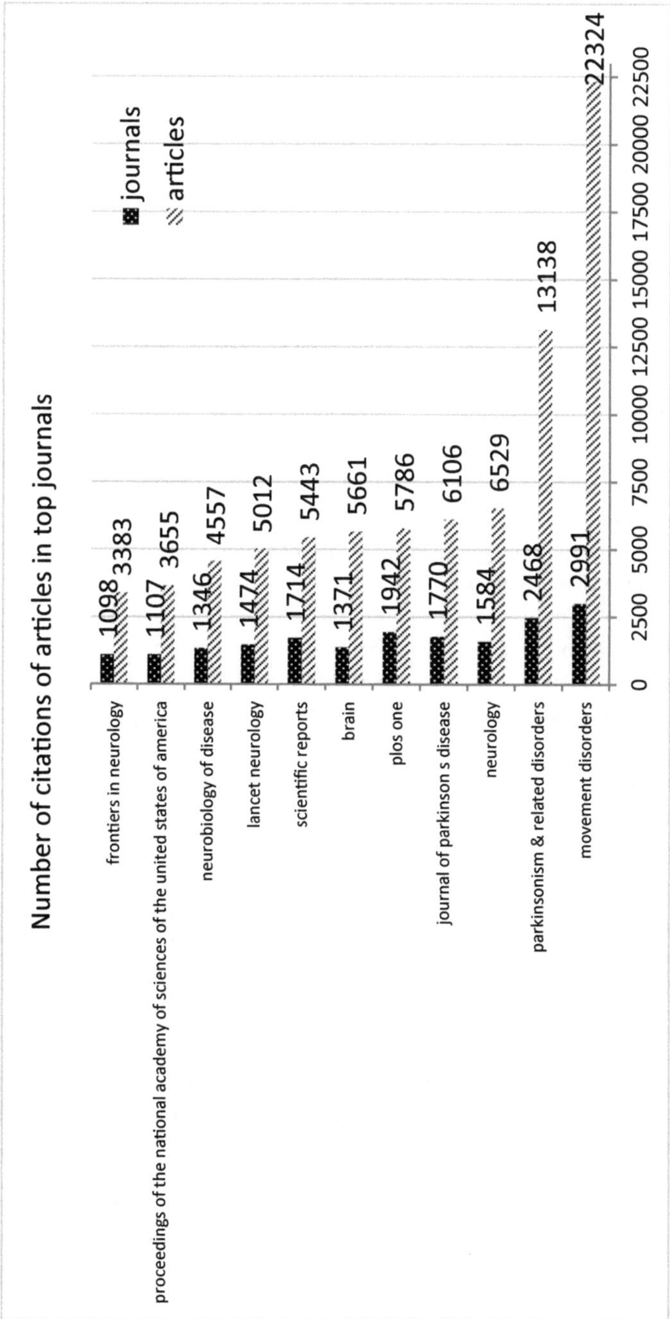

Fig. 6.8 Diagram of the number of articles and journals citing publications in the journal (2015–2021)

Table 6.2 List of journals with high citation rankings

№	Title of the journal	ISSN	Num of articles over 7 years	Number of citing journals
1	Movement Disorders	08853185 15318257	1201	2991
2	Parkinsonism & Related Disorders	18735126 13538020	1260	2468
3	PLOS One	19326203	670	1942
4	Journal of Parkinson's Disease	1877718X18777171	768	1770
5	Scientific Reports	20452322	696	1714
6	Neurology	00283878 1526632X	453	1584
7	Lancet Neurology	1474446514744422	80	1474
8	Brain	1460215600068950	252	1371
9	Neurobiology of Disease	1095953X09699961	349	1346
10	Journal of Neural Transmission	03009564 14351463	354	1297
11	Parkinson's Disease	20420080 20908083	445	1235
12	Journal of Neurochemistry	1471415900223042	202	1193
13	JAMA Neurology	21686157 21686149	96	1141
14	Proceedings of the National Academy of Sciences of the United States of America	10916490 00278424	176	1107
15	Frontiers in Neurology	16642295	678	1098
16	Molecular Neurobiology	08937648 15591182	335	1073
17	Journal of the Neurological Sciences	0022510X, 18785883	357	1071
18	Neuroscience Letters	1872797203043940	428	1044
19	Frontiers in Neuroscience	1662453X16624548	456	1042
20	Annals of Neurology	1531824903645134	133	1036
21	Journal of Neurology	0340535414321459	347	1027
22	npj Parkinson's Disease	23738057	253	1007
23	Nature Reviews Neurology	1759476617594758	74	1006
24	Frontiers in Aging Neuroscience	16634365	406	1003

The journal "Movement Disorders" was cited by 397 journals in their articles, and 2 journals cited it over 1000 times. These are "Parkinsonism & Related Disorders" (1135 citations) and the "Journal of Parkinson's Disease" (1074 citations). The number of journals citing "Movement Disorders" from 1 to 10 times is shown in the diagram (Fig. 6.9).

The journal "Parkinsonism & Related Disorders" was cited once by 438 journals and more than 500 times by three journals. These are the journals "Movement Disorders" (671 citations), "Frontiers in Neurology" (639 citations), and the "Journal of Parkinson's Disease" (558 citations).

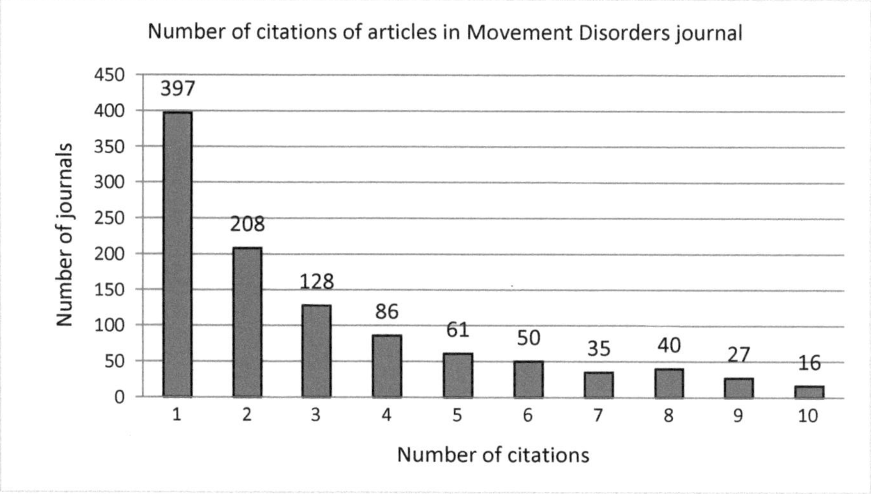

Fig. 6.9 Diagram of the number of journals citing the "Movement Disorders" journal up to 10 times

However, journals that do not rank highly by the number of publications can belong to the high-ranking citation list. For example, "Lancet Neurology," which ranks 83rd by publication count (80 articles on Parkinson's disease), ranks 7-th on the citation list. It was cited by 1,474 journals. As seen in Table 6.2, two other journals with fewer than 100 publications on Parkinson's disease over seven years also made this list. These are "JAMA Neurology," a peer-reviewed medical journal published by the American Medical Association (Q1; h-index 257; SJR 6.697; Impact Factor 20.4 (SCIE)), with 96 publications on Parkinson's disease over the past seven years, cited by 1,141 journals.

The journal "Nature Reviews Neurology" (Nature Publishing Group, SJR 6.536; Impact Factor 28.2 (SCIE)) published review articles on neurological research, in particular, 74 articles on Parkinson's disease from 2015–2021, which were cited by 1,006 journals. Thus, the number of citations a journal receives is determined not only by the number of publications in the journal but also by the diversity of researchers' interests.

The analysis of the number of citations per year showed a growth in citations year after year. Let us consider this using high-ranking journals as an example. Table 6.3 provides a list of journals that received more than 1,000 citations from 2015–2021, along with the number of citations each year.

A particular interest lies in journals where fewer than 100 articles on Parkinson's disease were published during the period under consideration. We can compare the number of publications in these journals distributed by year with the number of citations to the journals by year (Table 6.4).

Table 6.3 Citations to publications by year in high-ranking journals

№	Title of the journal	Num of articles over 7 years	Number of journals citing the journal per year						
			2015	2016	2017	2018	2019	2020	2021
1	Movement disorders	1201	111	277	387	432	514	603	667
2	Parkinsonism & related disorders	1260	57	188	316	364	422	532	589
3	PLOS One	670	22	138	252	288	368	418	456
4	Journal of Parkinson's Disease	768	27	96	196	247	318	384	502
5	Scientific Reports	696	25	62	194	255	338	381	459
6	Neurology	453	32	130	191	235	278	331	387
7	Lancet Neurology	80	22	98	136	183	250	356	429
8	Brain	252	41	97	169	203	236	292	333
9	Neurobiology of Disease	349	53	128	161	176	238	271	319
10	Journal of Neural Transmission	354	28	75	121	168	253	305	347
11	Parkinson's Disease	445	18	75	160	163	203	285	331
12	Journal of Neurochemistry	202	9	76	163	168	211	264	302
13	JAMA Neurology	96	32	83	129	157	194	255	291
14	Proceedings of the National Academy of Sciences of the United States of America	176	15	74	131	159	202	254	272
15	Frontiers in Neurology	678	4	17	49	95	197	332	404
16	Molecular Neurobiology	335	25	76	128	154	190	232	268
17	Journal of the Neurological Sciences	357	14	67	130	167	205	238	250
18	Neuroscience Letters	428	13	66	115	151	174	251	274
19	Frontiers in Neuroscience	456	5	30	89	114	189	273	342
20	Annals of Neurology	133	20	83	128	156	179	211	259
21	Journal of Neurology	347	26	67	109	125	182	222	296
22	npj Parkinson's Disease	253	2	24	72	120	202	274	313
23	Nature Reviews Neurology	74	19	55	116	141	163	230	282
24	Frontiers in Aging Neuroscience	406	6	38	83	128	185	253	310

As shown in Table 6.4, despite the small variation in the number of Parkinson's disease publications in these journals year by year, the interest in these publications grew. This resulted in a significant increase in the number of citations to the journal.

Next, we describe the results of the analysis of references in articles to publications in journals.

Table 6.4 Number of publications in the journal and the number of journals citing the journal, distributed by year

Journal	Number of publications / number of journals citing it in a year						
	2015	2016	2017	2018	2019	2020	2021
Lancet Neurology	11 /22	11/98	13/136	14/183	9/250	14/356	8/429
Jama Neurology	20/32	13/83	17/129	13/157	9/194	12/255	12/291
Nature Reviews Neurology	19/19	7/55	13/116	8/141	10/163	10/230	7/282

The citation analysis of articles showed that over the seven years under review, 567 journals had no references in the database, while 1 to 10 articles referred to 1,406 journals. More than 1,000 articles contained references to 59 journals, of which eight journals were referenced by more than 5,000 publications. Table 6.5 presents the list of these journals (Top-8).

Table 6.5 shows the total number of publications, including articles published in the journal itself, in which the journal is referenced. For comparison, the table provides the number of references to the journals excluding the articles published in the journal itself. As can be seen from the table, the leading journals in terms of the number of references in articles are the same as those in Table 6.2. In this list, the journals Parkinsonism & Related Disorders and Movement Disorders significantly outpace all other journals in terms of the number of references to their publications. It should also be noted that some journals have changed positions. For example, the journal PLOS One, which held third place in Table 6.2, moved to fifth place, and its place in Table 6.5 was taken by the journal Neurology, which received more references in articles. The journal Brain, which occupied the eighth position (Table 6.2), rose to sixth position in Table 6.5.

Table 6.5 Top-8 journals with the highest number of references to them in publications

№	Title of the journal	Number of citing articles	
		Total number of articles	excluding articles published in the journal itself
1	Movement Disorders	24122	22324
2	Parkinsonism & Related Disorders	14035	13138
3	Neurology	6851	6529
4	Journal of Parkinson s Disease	6524	6106
5	PLOS One	5911	5786
6	Brain	5909	5661
7	Scientific Reports	5613	5443
8	Lancet Neurology	5085	5012

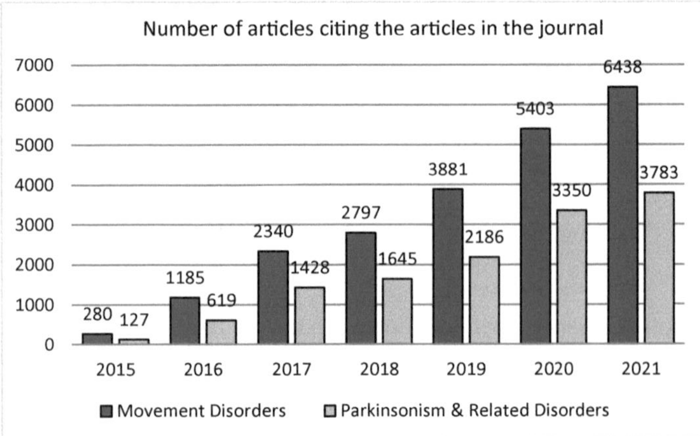

Fig. 6.10 Diagram of the number of references in articles to journals in each year under consideration

Comparing the total number of references and the number of references excluding publications in the journal itself, it can be observed that the rankings of journals do not change. Therefore, references to a journal included in the Top-8 in articles published in the journal itself do not raise the journal's ranking, i.e., do not increase its influence.

We also noted that the scientific article and preprint server in biology bioRxiv, which occupies third place in publication activity (795 publications on the topic of Parkinson's disease), did not enter the list of highly cited sources. Over the entire period, 466 articles and 206 journals cited publications in this archive.

An analysis of changes in the number of journal references by year was also conducted. Fig. 6.10 shows a diagram of the number of references in articles to the journals Parkinsonism & Related Disorders and Movement Disorders in each year under consideration. References in publications of the same journal were not counted.

A similar analysis was conducted for the journal Lancet Neurology. Fig. 6.11 shows the graph of changes in the number of references to the journal's publications on Parkinson's disease.

Let us consider the cross-references between high-ranking journals. Fig. 6.12 shows a diagram of the number of references from the journal Parkinsonism & Related Disorders to the journal Movement Disorders, and vice versa, from Movement Disorders to Parkinsonism & Related Disorders. The number of publications on PD during this period was given earlier in Fig. 6.3.

The diagram illustrates the influence of the journal Movement Disorders. The number of references to it from the journal Parkinsonism & Related Disorders generally exceeds the reverse references. At the same time, the ratio of the number of publications in the journals does not differ significantly. A similar situation is observed with references to the journal Movement Disorders from other journals. For instance,

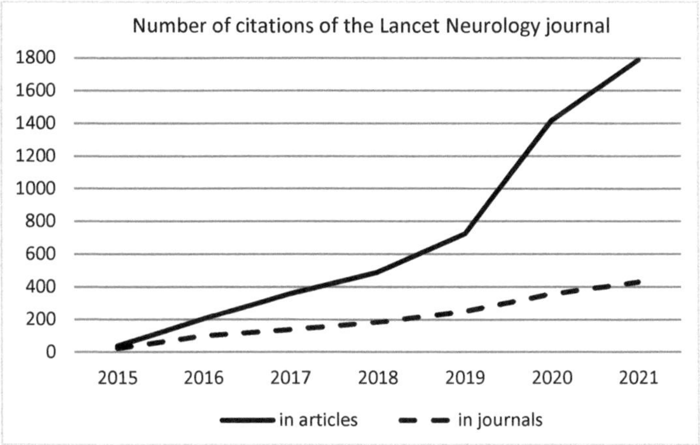

Fig. 6.11 Dynamics of changes in the number of references to the journal's publications

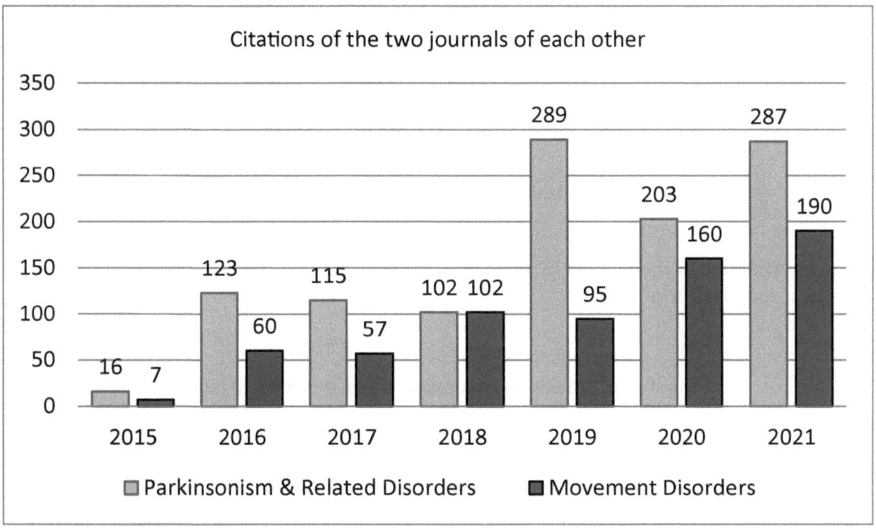

Fig. 6.12 Diagram of the number of references from the journal Parkinsonism & Related Disorders to the journal Movement Disorders, and vice versa

the Journal of Parkinson's Disease, which ranks fourth in the number of journal references to it (1,770 citing journals), provides 1,074 references in its publications to the journal Movement Disorders. In the same period, the journal Movement Disorders provides 265 references to Journal of Parkinson's Disease. The number of references to the journal Movement Disorders (24,122 articles refer to publications in this journal) indicates its influence in the field of publications on Parkinson's disease. However, some publications in other journals, which are less highly ranked in this research area, may attract specialists' attention, which is reflected in the number of references to the journal. Fig. 6.13 shows a diagram of yearly references from the journal Movement Disorders to the journal Brain and vice versa, from Brain to Movement Disorders. Let us compare the number of references with the number of publications on Parkinson's disease in the journals during this period. The publication diagram is presented in Fig. 6.14.

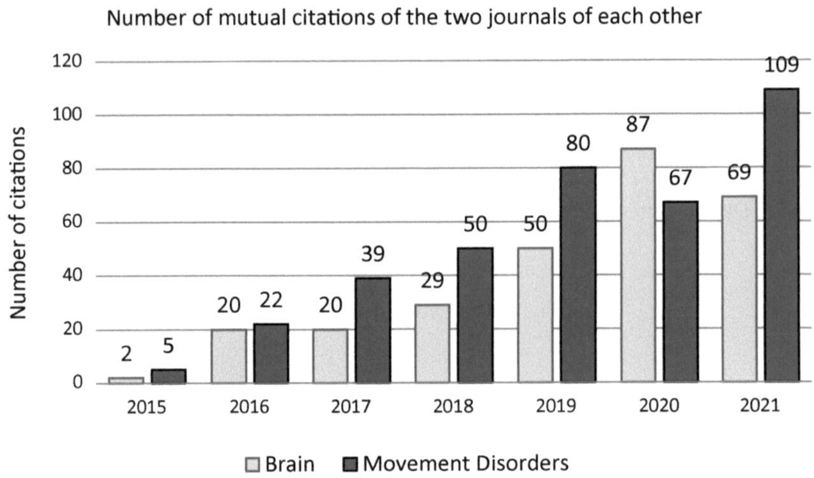

Fig. 6.13 Diagram of the number of references from the journal Brain to the journal Movement Disorders, and vice versa

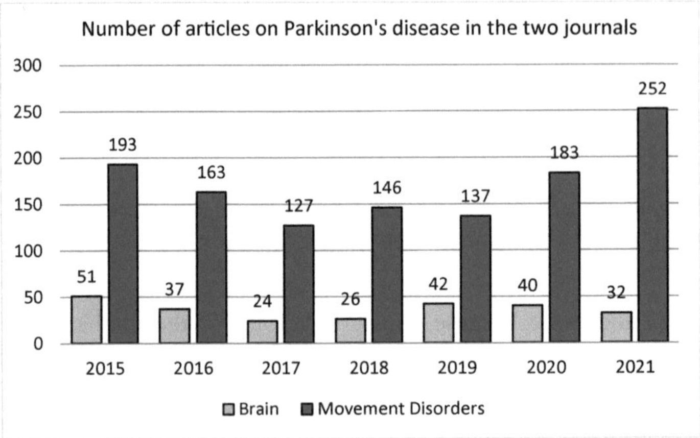

Fig. 6.14 Diagram of the number of publications in the journals Movement Disorders and Brain

The journal Brain publishes significantly fewer articles on Parkinson's disease, but these publications attract interest and receive a large number of references in a high-ranking journal.

In conclusion of the statistical review of the results of the study of publications on Parkinson's disease, we briefly describe journal "self-citation." In the downloaded database, 782 journals were identified in which references were made to publications in the same journal. In 2,510 journals, such references were absent.

Usually, authors choose proven high-ranking specialized journals for the publication of their articles. Many such journals are published by professional societies. Therefore, researchers often publish their articles in the same journal. Any scientific work is based on the experience of previous studies. These studies, as a rule, belong to other authors or develop and supplement the scientist's own research. Then, the articles contain references either to their own works or to the works of colleagues published in the same journal. A striking example of such a journal is "Movement Disorders," which is the journal of the International Parkinson and Movement Disorder Society (MDS). By the number of references to its own publications on PD (for 2015–2021), the journal ranks first—with 1,798 references. The journal "Parkinsonism & Related Disorders" is in second place with 897 references. The change in the number of self-citations by year for these journals is shown in Fig. 6.15 a, b.

In another 19 journals, more than 100 publications referred to articles published in the same journal. 288 journals over the entire period gave only one reference to their own publications, and in 107 journals there were two references to their own publications.

Thus, statistical analysis of the journal citation network allows for ranking journals by:

- number of publications;
- number of references to these publications in articles and journals;
- number of self-citations in their own publications.

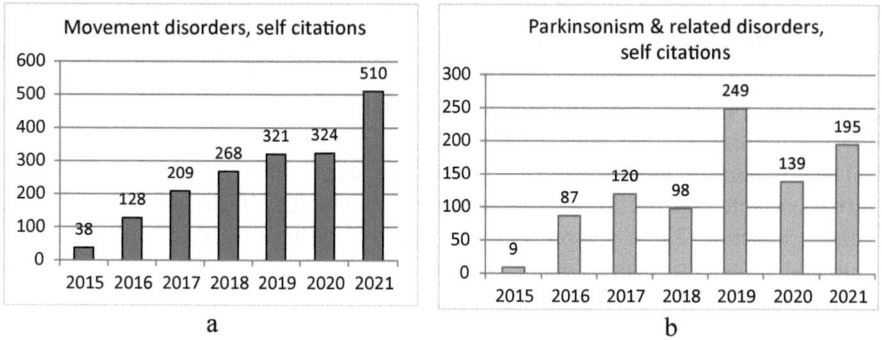

Fig. 6.15 Number of self-citations by year for the journals "Movement Disorders" (a) and "Parkinsonism & Related Disorders" (b)

It also allows for analyzing the dynamics of these parameters over time and assessing the degree of journal influence, including increase and decrease in impact.

Next, we consider the results of the analysis of network indices.

6.3 An Analysis of Journal Citation Network Indices

We calculated classical and new indices for the citation network by journals. The classical indices considered were In-degree, Eigenvector, Betweenness, and Page Rank, and the new indices were the Bundle Index and Pivotal Index. For the centrality analysis, the maximum size of the critical set was chosen as 3, and the following quotas were used (a quota is defined for each node separately as a percentage of the total number of references to each journal): 0.5%, 1%, 3%, 5%, 10%. The citation indices were calculated in two versions—with and without taking into account journal self-citation.

6.3.1 Classical Centrality Indices

When calculating the indices, journals were ranked, and each journal was assigned a rank depending on the value of the index. As an example, we present the In-degree index values for the Top-10 journals with and without self-citation (Table 6.6).

As seen from Table 6.6, the journals "Movement Disorders" and "Parkinsonism & Related Disorders" retain their leadership, even without accounting for self-citation. However, an analysis of the dynamics of the In-degree index by year for these journals showed a decline in the index (Fig. 6.16), despite maintaining leadership.

This may be due to several reasons, such as an increase in the number of publications on PD in other journals or a shift in researchers' interest to other research directions. An example of index growth is shown in Fig. 6.17. The journal "Scientific Reports" (Nature Publishing Group) ranked 19th in our rating in 2015 and moved to 26th in 2016. However, in 2018 it entered the Top-10, taking 7th position. Then, during 2019–2021, it continued to be in the Top-10, occupying the 5th–6th positions. The journal "Frontiers in Neuroscience" was in 93rd place in 2015–2016, and in 2021 it moved to 9th place and ranked 17th overall.

The comparative analysis of the classical indices considered by us for the Top-10 journals selected by the In-degree index showed changes in the rankings of some of these journals. Table 6.7 presents the Top-10 journals by the In-degree index value and shows the journal ranks by other considered indices taking into account self-citation.

As can be seen from Table 6.7, only four journals remain in the Top-10 across all indices. The journals "Movement Disorders" and "Parkinsonism & Related Disorders" maintain their leadership across all indices. A huge number of citations and publications explain the fact that these two journals are the best in all indices. The

Table 6.6 Top-10 journals by In-degree index with and without self-citation

ISSN	Title of the journals	In-degree with self-citations	Rank	In-degree without self-citations	Rank
08853185, 15318257	Movement Disorders	0.0776	1	0.0746	1
18735126, 13538020	Parkinsonism & Related Disorders	0.0451	2	0.0439	2
00283878, 1526632X	Neurology	0.022	3	0.0218	3
1877718X18777171	Journal of Parkinson s Disease	0.0209	4	0.0204	4
19326203	PLOS One	0.019	5	0.0193	5
14602156, 00068950	Brain	0.019	6	0.0189	6
20452322	Scientific Reports	0.018	7	0.0182	7
14744465, 14744422	Lancet Neurology	0.0163	8	0.0167	8
1095953X09699961	Neurobiology of Disease	0.0153	9	0.0152	9
10916490, 00278424	Proceedings of the National Academy of Sciences of the United States of America	0.012	10	0.0122	10

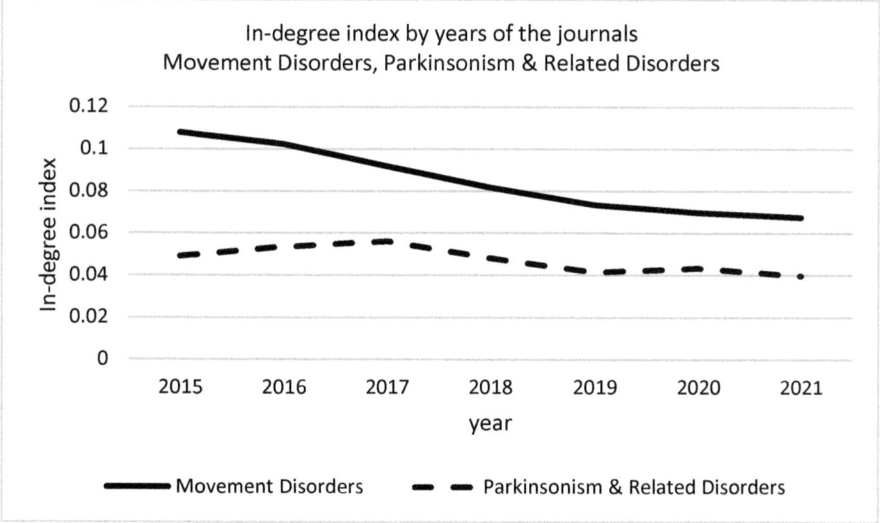

Fig. 6.16 Dynamics of the In-degree index by year for the journals "Movement Disorders" and "Parkinsonism & Related Disorders"

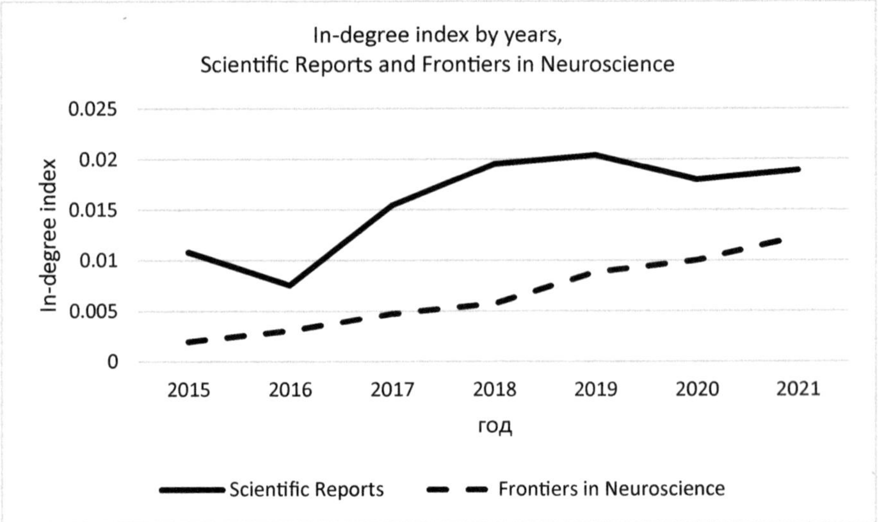

Fig. 6.17 Dynamics of the In-degree index by year for the journals "Scientific Reports" and "Frontiers in Neuroscience"

Table 6.7 Top-10 journals by In-degree index

Title of the journals	In-degree values	Rank			
		In-degree	Betweenness	Eigenvector	PageRank
Movement Disorders	0.0776	1	1	1	1
Parkinsonism & Related Disorders	0.0451	2	2	2	2
Neurology	0.022	3	15	3	3
Journal of Parkinson s Disease	0.0209	4	3	5	4
PLOS One	0.019	5	4	11	5
Brain	0.019	6	25	4	7
Scientific Reports	0.018	7	5	10	8
Lancet Neurology	0.0163	8	19	6	6
Neurobiology of Disease	0.0153	9	17	13	10
Proceedings of the National Academy of Sciences of the United States of America	0.012	10	48	16	15

journals "Journal of Parkinson's Disease" and "Scientific Reports" are also in the Top-10 across all indices.

The Top-10 journals by the Betweenness index (Table 6.8) differ in the last five positions (the journals "Frontiers in Neurology", "International Journal of Molecular Sciences", "Parkinson's Disease", "Frontiers in Neuroscience", "Frontiers in Aging Neuroscience"). This is due to the fact that they do not have a large number of citations but are often found in networks on paths from one journal to another and are important "transmitters of information".

The dynamics of the Betweenness index over the years for two high-ranking journals is shown in Fig. 6.18.

The index of the first journal "Movement Disorders", which is the most cited, has a clear downward trend and has decreased by half over the entire period. The index of the journal "Parkinsonism & Related Disorders" has practically not changed.

The downward trend in indices may be related to the fact that publications in recent years are still cited less than older publications from 2015–2017. The journal "International Journal of Molecular Sciences" (ISSN 1661-6596 for the print edition), which had a rank of 27 by In-degree, entered the Top-10 journals by the Betweenness index. The Betweenness index of this journal, as well as that of the "Journal of Parkinson's Disease" (rank 3 by In-degree), had an upward trend during the considered time period (Fig. 6.19). In 2021, the "Journal of Parkinson's Disease" became more influential.

Of the Top-10 journals by Eigenvector centrality, 7 journals are among the ten most cited journals (Table 6.9). The new journals that entered the Top-10 are two peer-reviewed American journals: "Annals of Neurology" (publisher John Wiley & Sons Inc. USA, ISSN 15318249, 03645134), and "JAMA Neurology" (published by

Table 6.8 Top-10 journals by Betweenness

Title	Betweenness values	Rank		
		In- degree	Eigenvector	PageRank
Movement Disorders	0.0645	1	1	1
Parkinsonism & Related Disorders	0.0608	2	2	2
Journal of Parkinson's Disease	0.048	4	5	4
PLOS One	0.0396	5	11	5
Scientific Reports	0.036	7	10	8
Frontiers in Neurology	0.0294	11	7	9
International Journal of Molecular Sciences	0.0274	34	48	37
Parkinson's Disease	0.026	21	24	11
Frontiers in Neuroscience	0.0249	17	26	21
Frontiers in Aging Neuroscience	0.0215	27	25	25

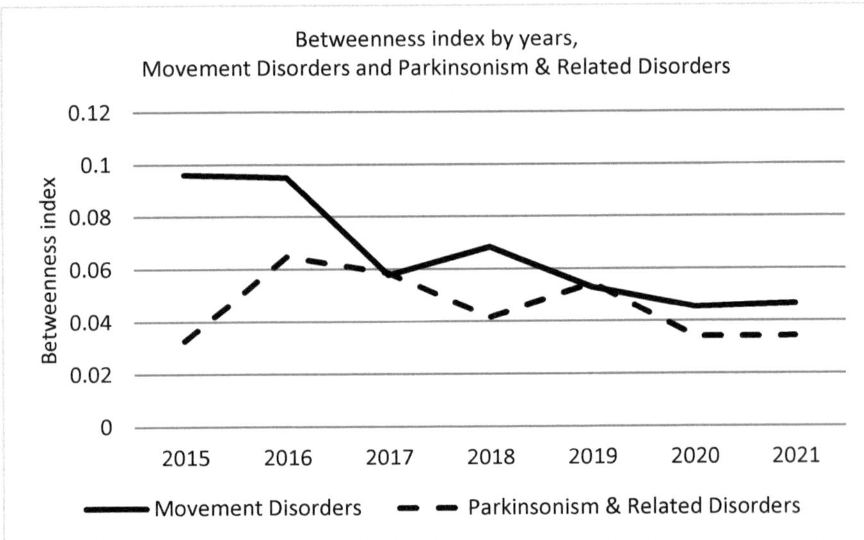

Fig. 6.18 Dynamics of the Betweenness index over the years for the journals "Movement Disorders" and "Parkinsonism & Related Disorders"

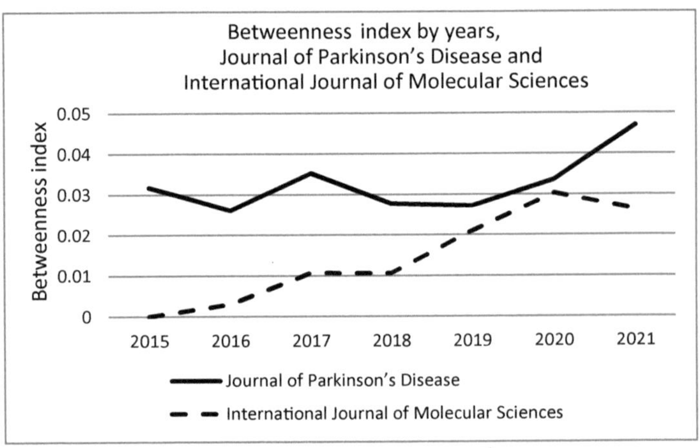

Fig. 6.19 Dynamics of the Betweenness index over the years for the journals "Journal of Parkinson's Disease" and "International Journal of Molecular Sciences"

the American Medical Association, USA, ISSN 21686211). The list also includes the journal "Frontiers in Neurology" (published by Frontiers Media S.A., ISSN 1664-2295), which also entered the Top-10 by the Betweenness index.

Thus, the journals "Movement Disorders" and "Parkinsonism & Related Disorders" are also the most important in the field of PD publications according to the

Table 6.9 Top-10 journals by Eigenvector

№	Journal	Eigenvector values	Rank		
			In-degree	Betweenness	PageRank
1	Movement Disorders	0.7599	1	1	1
2	Parkinsonism&Related Disorders	0.3581	2	2	2
3	Neurology	0.2089	3	15	3
4	Brain	0.1806	6	25	7
5	Journal of Parkinson's Disease	0.1694	4	3	4
6	Lancet Neurology	0.1610	8	19	6
7	Frontiers in Neurology	0.1122	11	6	9
8	Annals of Neurology	0.1043	13	45	14
9	JAMA Neurology	0.1041	12	65	12
10	Scientific Reports	0.0947	7	5	8

Eigenvector and maintain their leadership. The analysis of changes in the Eigenvector index for these journals over the years (Fig. 6.20) showed minor fluctuations in the index. It is worth noting that the index of the journal "Movement Disorders" had the highest values over 7 years with a huge lead over the other journals.

It is interesting to compare the changes in the Eigenvector index for the new journals that entered the Top-10: "Annals of Neurology" and "JAMA Neurology"

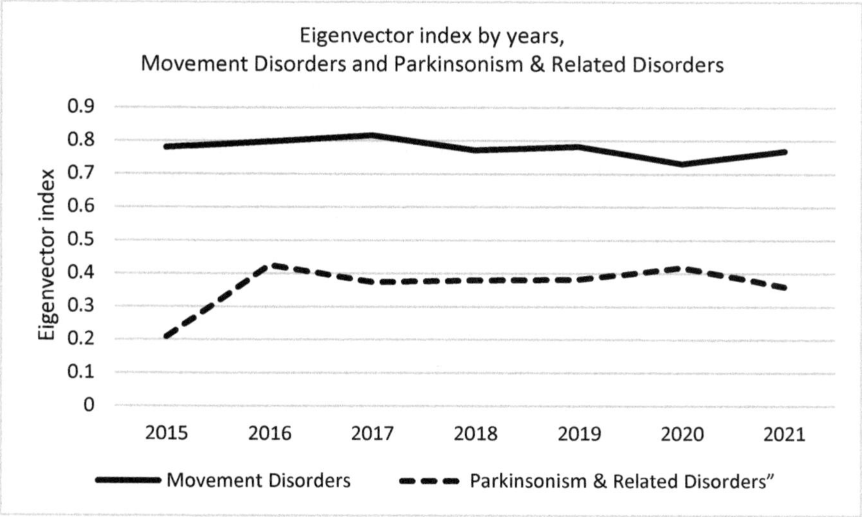

Fig. 6.20 Eigenvector index over the years for the journals "Movement Disorders" and "Parkinsonism & Related Disorders"

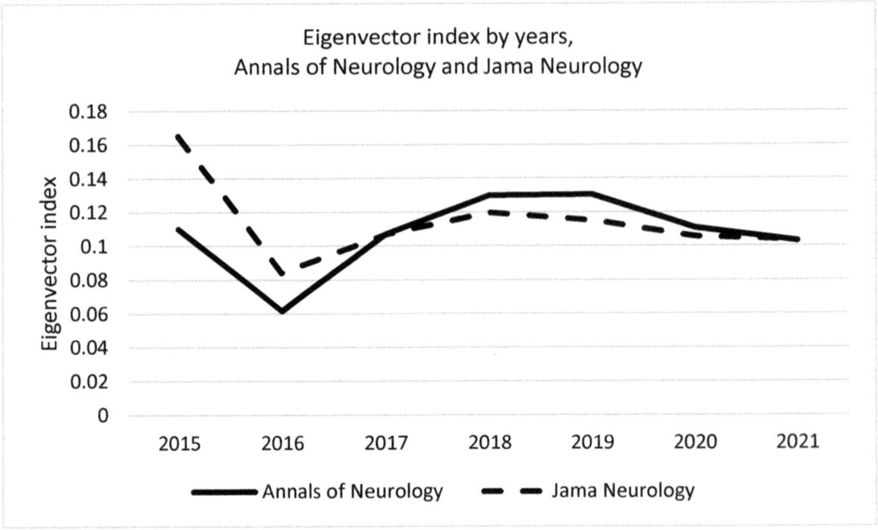

Fig. 6.21 Dynamics of the Eigenvector index over the years for the journals "Annals of Neurology" and "JAMA Neurology"

(Table 6.9). The dynamics of the index for these journals over the years is shown in Fig. 6.21.

It is worth noting that in 2015 the journal "JAMA Neurology" had rank 4 by the Eigenvector index, and "Annals of Neurology" had rank 7. Even after some drop in ranks in 2016 ("JAMA Neurology"—rank 8, "Annals of Neurology"—rank 13), in the following years the journals maintained their positions in the Top-10, i.e., the journals had considerable weight during this period.

And the last classical index in our comparative analysis is PageRank. The Top-10 journals ranked by the PageRank index are presented in Table 6.10. The results for PageRank, with minor exceptions, are similar to the results of In-degree. The same highly cited journals are listed with slight rearrangements and the addition of the journal "Frontiers in Neurology" to the top ten. The table also presents ranks by other considered indices.

The conducted comparative analysis of the four classical centrality indices made it possible to identify the most important and influential journals out of the 3292 journals considered.

However, as we have already indicated, classical indices do not allow accounting for group and indirect influence of some nodes on others, therefore, new indices were considered—BI (Bundle Index)—the index of group influence; PI (Pivotal Index)—the index of influence through pivotal nodes; as well as Total Influence (TI)—the index of overall influence.

Table 6.10 Top-10 journals by PageRank

№	Journals	PageRank values	Rank		
			In-degree	Betweenness	Eigenvector
1	Movement Disorders	0.044	1	1	1
2	Parkinsonism & Related Disorders	0.0288	2	2	2
3	Neurology	0.0144	3	15	3
4	Journal of Parkinson's Disease	0.0143	4	3	5
5	PLOS One	0.0129	5	4	11
6	Lancet Neurology	0.0128	8	19	6
7	Brain	0.0115	6	25	4
8	Scientific Reports	0.0109	7	5	10
9	Frontiers Neurology	0.0086	11	6	7
10	Neurobiology Disease	0.0082	9	17	13

6.3.2 New Centrality Indices

New centrality indices were calculated, as well as overall influence, by which journals were ranked. The calculation results showed that the Top-10 journal list for the quotas 0.5%, 1%, and 3% is practically identical to the list obtained for classical indices. As an example for the Top-10 journals, the value of the TI index and the ranks of the BI and PI indices for the quota q=0.5% are provided. (Table 6.11). The calculations were carried out taking into account journal self-citation. For comparison, the values of the journals' In-degree are given. The journals are ranked by the In-degree index.

As seen from Table 6.11, according to the total influence index (TI), two journals (No. 9 and No. 10) fell outside the Top-10 most cited journals. The Top-10 by the TI index included "Journal of Neural Transmission" (ISSN: 03009564, 14351463; Publisher: Springer Verlag), which took the 9th position, and "Parkinson's Disease" (ISSN: 20908083, 20420080; Publisher: Hindawi, London), which has the 10th rank.

The Top-10 journals for quotas q=1% and q=3% are presented in Table 6.12 and Table 6.13, respectively. The journals are ranked by TI. The tables show the ranks of the journals for BI and PI.

The Top-10 lists for the quotas 0.5%, 1%, and 3% feature the same journals. Analysis of the index values for the quotas of 5% and 10% revealed some changes in the ranked lists of journals. Thus, in the Top-10 journals with a quota of 5%, three journals appeared that had not previously been in the Top-10 by In-degree values. Table 6.14 presents the Top-10 list for the 5% quota. The table provides the values of the In-degree, TI, BI, and PI indices. The list is ranked by TI.

The leading journals still maintain their positions. However, the following journals entered the top:

Table 6.11 Values of TI and In-degree indices and ranks of TI, BI, and PI (q=0.5%) for the Top-10 journals

№	Journal	In-degree values	Rank q=0.5%			
			TI	TI	BI	PI
1	Movement Disorders	0.0776	0.0632	1	1	1
2	Parkinsonism & Related Disorders	0.0451	0.0444	2	2	2
3	Neurology	0.0220	0.0205	6	7	10
4	Journal of Parkinson's Disease	0.0210	0.0273	4	4	4
5	Plos One	0.0190	0.0312	3	3	3
6	Brain	0.0190	0.0174	8	11	13
7	Scientific Reports	0.0181	0.0245	5	5	5
8	Lancet Neurology	0.0164	0.0204	7	6	6
9	Neurobiology of Disease	0.0153	0.0163	11	12	12
10	Proceedings of the National Academy of Sciences of the United States of America	0.0121	0.0121	20	22	24

Table 6.12 Top-10 journals by TI, q=1%

Ранг TI	Journals	Rank q=1%	
		BI	PI
1	Movement Disorders	1	1
2	Parkinsonism & Related Disorders	2	2
3	PLOS One	3	3
4	Journal of Parkinson's Disease	4	4
5	Neurology	5	6
6	Lancet Neurology	6	5
7	Scientific Reports	7	7
8	Brain	9	12
9	Neurobiology of Disease	11	8
10	Journal of Neural Transmission	8	10

- JAMA Neurology (ISSN: 21686149, Publisher: American Medical Association: 6th position in the top),
- Journal of Neurology (ISSN: 03405354, 14321459, published by the D. Steinkopff-Verlag: 8th position in the top),
- At No. 10 in the list is "Journal of Neural Transmission", which entered the Top-10 already at a quota of 0.5% (see Table 6.11).

The Top-10 journals for the 10% quota are presented in Table 6.15. The table gives the values of the In-degree, TI, BI, and PI indices and their ranks. The list is ranked by TI.

Table 6.13 Top-10 journals by TI, q=3%

Rank TI	Journals	Rank q=3%	
		BI	PI
1	Movement Disorders	1	1
2	Parkinsonism & Related Disorders	2	2
3	Journal of Parkinson's Disease	3	3
4	Neurology	5	6
5	PLOS One	4	5
6	Scientific Reports	6	4
7	Lancet Neurology	7	7
8	Brain	9	9
9	Journal of Neural Transmission	8	8
10	Neurobiology of Disease	10	11

Table 6.14 Top-10 journals by TI, q=5%

№	Title	Values				Rank			
		In-degree	TI	BI	PI	In-degree	TI	BI	PI
1	Movement Disorders	0.0776	0.0575	0.0533	0.0417	1	1	1	1
2	Parkinsonism & Related Disorders	0.0452	0.0423	0.0451	0.0367	2	2	2	2
3	Neurology	0.022	0.0292	0.037	0.0287	3	3	3	3
4	Journal of Parkinson's Disease	0.021	0.0205	0.0235	0.0171	4	4	4	5
5	Lancet Neurology	0.0164	0.017	0.0192	0.0153	8	5	5	6
6	JAMA Neurology	0.0109	0.0149	0.0168	0.0171	12	6	7	4
7	Brain	0.019	0.0144	0.0134	0.0108	6	7	10	9
8	Journal of Neurology	0.0108	0.0137	0.0175	0.0134	16	8	6	7
9	Neurobiology of Disease	0.0153	0.012	0.0086	0.012	9	9	16	8
10	Journal of Neural Transmission	0.0108	0.0115	0.0138	0.01	14	10	9	11

Table 6.15 Top-10 journals by TI, q=10%

№	Title	Values				Rank			
		In-degree	TI	BI	PI	In-degree	TI	BI	PI
1	International Journal Of Molecular Science	0.0063	0.0409	0.0730	0.0434	33	1	1	1
2	Neurological Sciences	0.0086	0.0278	0.0494	0.0293	22	2	2	2
3	Movement Disorders	0.0776	0.0278	0.0025	0.0030	1	3	24	25
4	Parkinsonism & Related Disorders	0.0452	0.0167	0.0022	0.0027	2	4	30	30
5	Elife	0.0042	0.0143	0.0243	0.0144	47	5	3	4
6	Neurology	0.0220	0.0095	0.0052	0.0035	3	6	14	16
7	Neurotoxicity Research	0.0020	0.0091	0.0107	0.0145	82	7	5	3
8	The Embo Journal	0.0019	0.0091	0.0114	0.0138	91	8	4	5
9	Journal of Parkinson's Disease	0.0210	0.0079	0.0011	0.0014	4	9	164	181
10	Brain	0.0190	0.0070	0.0009	0.0011	6	10	262	276

As can be seen from Table 6.15, the Top-10 journal list changed drastically. The highly cited journals "Movement Disorders" and "Parkinsonism & Related Disorders", which occupied the first and second positions in the ranking of all considered classical indices, dropped below the Top-20. The journals "Journal of Parkinson's Disease" and "Brain", which were also in the top ten, fell outside the top hundred. The leaders became:

"International Journal of Molecular Science"—(ISSN: 14220067, 16616596; published by the Multidisciplinary Digital Publishing Institute (MDPI), Switzerland. New journal, founded in 2000. Q1; h-index—230; Impact Factor—6.208);

"Neurological Sciences"—(ISSN: 15901874, 15903478, published by Springer-Verlag Italia Srl. Q2; h-index—83; Impact Factor—3.83);

Journal "Elife", publishing articles on biochemistry, genetics, and molecular biology. (ISSN: 050084X; publisher: eLife Sciences Publications, UK; Q1; h-index—161; Impact Factor—7.937);

Journal "Neurotoxicity Research", publishing studies related to neurobiology, pharmacology, and toxicology. (ISSN: 14763524, 10298428; Publisher: Springer New York. Q2; h-index—81; Impact Factor—3.978);

"The EMBO Journal", specializing in publications on molecular research (ISSN: 02614189, 14602075; Publisher: The European Molecular Biology Organization. Q1; h-index—405; Impact Factor—10.859)

The bibliometric characteristics of the journals are provided as of 2021.

Thus, increasing the quota allowed highlighting highly specialized journals publishing research in molecular biology and medicine, genetics, and biochemistry. And although in terms of the total number of citations they lag behind broad-profile specialized journals, they are leaders in citation within their thematic scope. This effect was noted in Chapter 2 when defining the Bundle Index and Pivotal Index.

6.4 Clustering of Journal Patterns by Change in Their Citation by Years

An analysis of patterns with various parameters was conducted for 3,292 journals. The main task of such analysis is clustering of journals and identification of differing patterns. This made it possible to determine the journals whose citation increased during the period from 2015 to 2021, and those journals whose citations declined. Pattern analysis was carried out based on changes in the In-degree index by year. For the analysis, an epsilon tube of $\varepsilon = 0.05$ was selected. Two variants of clustering were performed—clustering by absolute value and by slope angle.

6.4.1 Clustering by Absolute Value

Clustering by absolute value makes it possible to determine journals whose number of citations differs significantly from the citation of the majority of journals. In this variant, 20 patterns were identified. The largest (first) pattern included 3,228 out of 3,292 journals: the citation values were close to zero for all 7 years.

The second pattern included 12 journals. The In-degree index of these journals either grew slightly or remained unchanged over the 7 years. Of the 12 journals, 11 did not enter the Top-10 by any of the indices considered.

The third pattern included 10 journals. Their index values did not change over the considered period, and the values were also quite low; none of the journals entered the Top-10.

The fourth pattern included 9 journals. In 2015–2017 they had an almost zero In-degree index, but by 2021 it had increased. The journals in this pattern did not enter the Top-10 list by any index considered. However, this pattern combined 5 journals publishing research related to biochemistry, genetics, and molecular biology ("ACS Chemical Neuroscience", ISSN: 19487193; "Cell Reports", ISSN: 22111247; "Cells", ISSN: 20734409; "Stem Cell Reports", ISSN: 22136711; "Frontiers in Molecular Neuroscience" ISSN: 16625099). For example, the journal "Stem Cell

Reports" was cited by 4 journals (5 articles) in 2015, and by 101 journals (273 articles) in 2021. The number of journal citations to "Cell Reports" increased from 11 in 2015 to 153 in 2021. This pattern also included two journals on translational research ("Annals of Clinical and Translational Neurology", ISSN: 23289503; and "Science Translational Medicine", ISSN: 19466234), a journal on neurobiology ("International Review Of Neurobiology", ISSN:747742), and a journal on immunology and microbiology ("Journal of Neuroinflammation", ISSN:17422094). One can assume an increased interest in the subject matter of the journals included in this pattern.

The fifth pattern also included 9 journals. The In-degree index values of these journals increased over the 7 years, but did not exceed 0.1. Interestingly, this pattern contains far fewer specialized neurological journals; there are general medical journals.

The sixth pattern included 4 journals, whose In-degree values were high in 2015 and declined by 2021. All journals specialize in neurology.

"Neurobiology of Aging" (ISSN: 01974580, 15581497; published by Elsevier Inc.; Q1; h-index—205; Impact Factor—5.133). Specializes in age-related diseases.

"Neuroscience" (ISSN: 03064522, 18737544; published by Elsevier Inc.; Q2; h-index—238; Impact Factor—5.133). Publishes research related to neuroscience.

"Journal of Neurology, Neurosurgery and Psychiatry" (ISSN: 1468330X, 00223050; BMJ Publishing Group, United Kingdom; Q1; h-index—225).

"JAMA Neurology" (ISSN: 21686157, 21686149; published by the American Medical Association; Q1; h-index—257; Impact Factor—29.907). This journal entered the Top-10 by Eigenvector index. Its in-degree initially fell in 2016, but later rose again.

Figure 6.22 shows the dynamics of the In-degree index by year for these journals belonging to the 6th pattern.

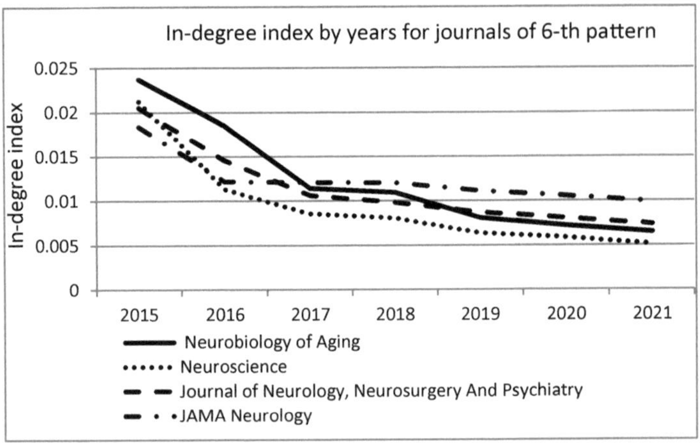

Fig. 6.22 Dynamics of In-degree index by years for journals of the 6-th pattern

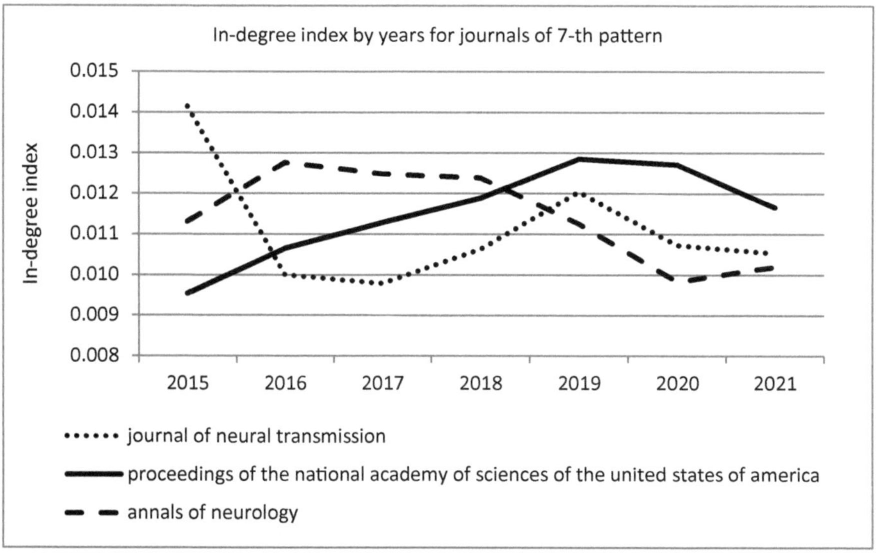

Fig. 6.23 Dynamics of the In-degree index by years for journals of pattern 7

The 7-th pattern included 3 journals (Fig. 6.23). Their index had more pronounced changes and in absolute values exceeded the index values for journals from the first cluster. These are the journals: "Journal of Neural Transmission" (ISSN: 03009564, 14351463; published by Springer Verlag; Q2; h-index—124; Impact Factor—3.85); "Proceedings of the National Academy of Sciences of the United States of America" (ISSN: 10916490, 00278424; Q1; h-index—838; Impact Factor—12.779). The third journal, "Annals of Neurology" (ISSN: 15318249, 03645134; published by John Wiley & Sons Inc., USA; Q1; h-index—321; Impact Factor—11.274) ranked 8th by Eigenvector index.

In Pattern 8 there are 3 journals: "Frontiers in Neuroscience", "Frontiers in Aging Neuroscience", and "npj Parkinson's Disease". Since 2016, these journals have shown a trend of increasing citation.

The following patterns included 2 journals each.

Thus, in Pattern 9 there are 2 journals—"Nature Neuroscience" and "Acta Neuropathologica". Both journals specialize in neurology, the index values are small, and there is also a slight increase in the In-degree index.

Pattern 10 included two journals: "International Journal of Molecular Sciences" and "Molecular Neurodegeneration"—a slight increase in the In-degree index is observed. The journals do not appear in the Top-10; interestingly, both specialize in molecular biology.

The next patterns included one journal each; as a rule, these journals have the highest indices.

Pattern 11 includes one journal—"Journal of Parkinson's Disease", its In-degree increased every year, and the journal frequently appeared in the Top-10.

Pattern 12 includes the journal "Brain", which appeared in the Top-10 according to classical indices (except for Betweenness); its citation level remained approximately the same.

Pattern 13 includes the journal "Neurobiology of Disease"—ranked 10th in 4 indices.

Pattern 14 includes the journal "Scientific Reports"—its citation has increased since 2016, it often appeared in the Top-10 by various indices; however, since 2019, the citation growth has slowed down.

Two journals occupying top positions by all indices—"Movement Disorders" and "Parkinsonism & Related Disorders" fell into patterns 15 and 18, respectively.

Pattern 16 includes the journal "Lancet Neurology"—also often in the Top-10; citation increased from 2015 to 2021.

Pattern 17 includes "PLOS One"—has high citation, the index increased until 2018, from 2018 to 2021 a decrease in the index is observed. This journal also appeared in the Top-10 by various indices.

Pattern 19 includes one of the most popular journals—"Neurology". The citation index of the journal during the studied period ranged between 0.0297 and 0.0222.

Pattern 20 includes the journal "Frontiers in Neurology". Since 2018, a significant increase in citation is observed.

6.4.2 Clustering by Slope Angle

This method allows identifying journals similar in dynamic changes of citation over 7 years. Clustering by the tangents of the broken line slope angles was also performed with the parameter $\varepsilon = 0.05$. Ten patterns were identified. Pattern 1 included 3281 out of 3292 journals. These journals did not show a clear trend over the 7 years. They maintained their index values.

The remaining groups included 1–2 journals each:

Pattern 2 included 2 journals: "Journal of Neurochemistry" and "Parkinsonism & Related Disorders"—the second is a highly cited journal. The first journal has low citation indicators, therefore it did not appear in the Top-10. The citation of these journals does not increase by 2021.

Pattern 3 also includes 2 journals: "Neurobiology of Aging" and "Neuroscience". They show a trend of decreasing In-degree index, and they also stood out as a single pattern by absolute value.

All the following patterns (4 to 9) include journals that also stood out in separate patterns by absolute value. Generally, all these journals have rather high In-degree index values and appeared in the Top-10 journals by various indices. These journals are "PLOS One" (Fig. 5.24a), "Brain" (Fig. 5.24b), "Frontiers in Neurology" (Fig. 5.24c), "Neurobiology of Disease" (Fig. 5.24d), "Frontiers in Neuroscience" (Fig. 5.24e), and "Lancet Neurology" (Fig. 5.24f).

The presented graphs reflect the nature of changes in journal citation over different periods. Thus, the citation of the journal "PLOS One" (Fig. 6.24a) increased until

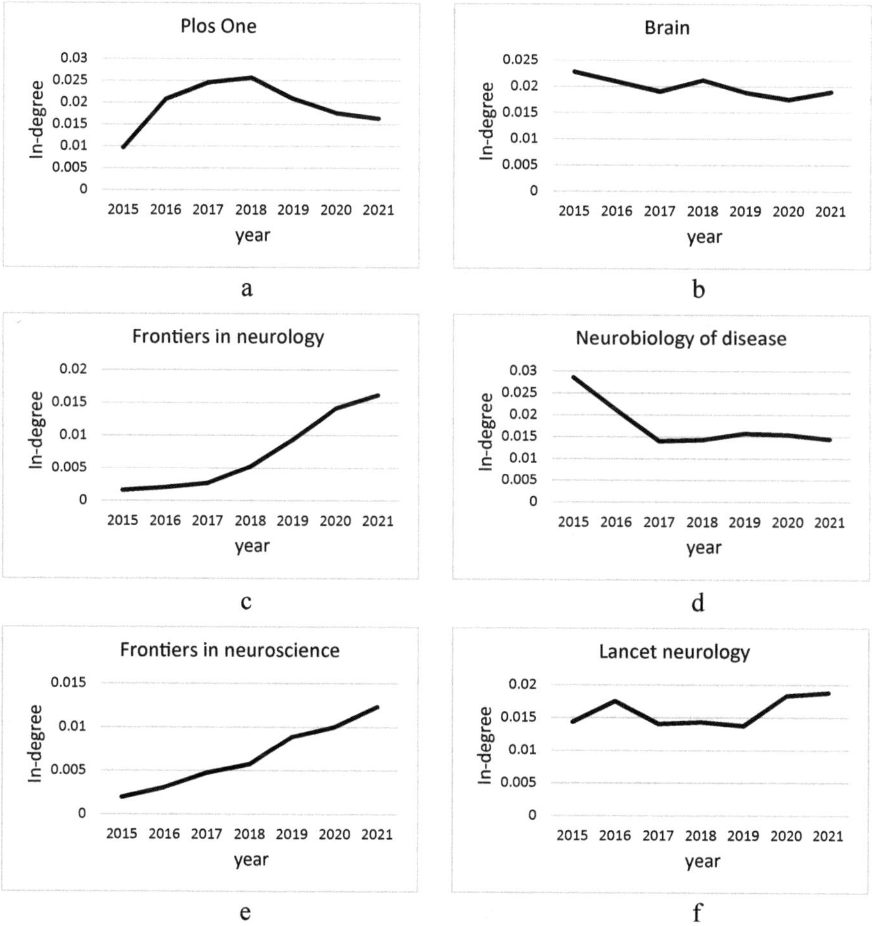

Fig. 6.24 In-degree value for journals included in patterns 4 (a), 5 (b), 6 (c), 7 (d), 8 (e), 9 (f)

2018 and then began to decline. The journals "Frontiers in Neurology" (Fig. 6.24c) and "Frontiers in Neuroscience" (Fig. 6.24e), on the contrary, showed an increase in citation over the studied time period.

Pattern 10 included the journal "Scientific Reports". In the pattern analysis by absolute values, it also stood out in a separate cluster. Its citation began to increase in 2016, continued to grow until 2019, and then remained unchanged.

Thus, this method makes it possible to identify journals whose dynamic changes in citation differ from the majority of journals in terms of absolute values.

6.5 An Analysis of Journal Citation Network Stability

A stability analysis was conducted for the journal citation network consisting of 3,292 nodes. 6 networks were constructed for each year from 2015 to 2021, evaluating the dynamic changes in the citation network. The in-degree index was chosen as the centrality indicator of a node for a specific year. The coefficient of influence of one node on another was calculated as follows:

$$\tilde{c}_{ij}^{t} = \begin{cases} \frac{w_{ij}}{\sum_k w_{kj}}, & if\ an\ edge\ from\ i\ to\ j\ exists \\ 0, & else \end{cases}$$

The results of the stability analysis can be represented on a two-dimensional plane (Fig. 6.25). The horizontal axis indicates the similarity of the networks in terms of centrality, and the vertical axis indicates similarity in terms of structure.

It should be noted that for the journal citation network, changes in centrality are of greater importance, since over seven years, changes in the structure of such a network are almost negligible. The greatest changes in centrality occurred from 2015 to 2016, and the greatest changes in network topology occurred from 2020 to 2021. We introduce a general measure of network similarity as the length of a vector from the origin to the corresponding point on the graph. By measuring the network stability in this way, it can be concluded that the greatest changes in the network occurred from 2015 to 2016. Thereafter, the network became increasingly stable each year. The similarity measures of the networks are presented in Table 6.16.

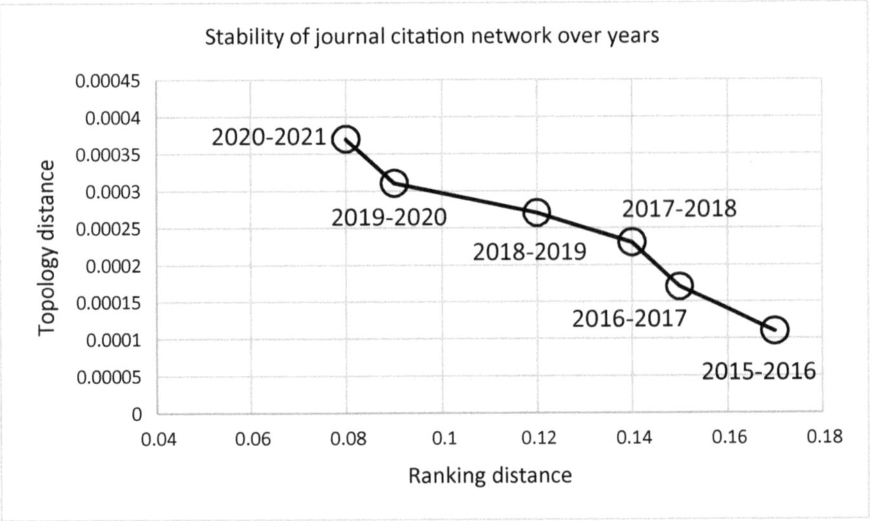

Fig. 6.25 Stability of the journal citation network from 2015 to 2021

Table 6.16 Stability of the journal citation network from 2015 to 2021

Year	Ranking distance	Topology distance	Length of vector
2015–2016	0.1771	0.00011	0.1252
2016–2017	0.1514	0.00018	0.1071
2017–2018	0.1338	0.00023	0.0946
2018–2019	0.1181	0.00027	0.0835
2019–2020	0.0975	0.00032	0.0689
2020–2021	0.0913	0.00036	0.0646

6.6 Main Findings from the Analysis of Journal Citations

The analysis of 3,292 journals allowed us to:

1. Rank journals by the number of publications on PD from 2015 to 2021. The journals with the highest number of publications are Parkinsonism & Related Disorders (1,260 publications) and Movement Disorders (1,201 publications);
2. Track the dynamics of changes in publication activity of each journal on the given topic over the seven-year period. Identify journals with increasing and decreasing numbers of publications. For example, in the journal "Frontiers in Aging Neuroscience", the number of publications on PD increased from 23 in 2015 to 108 in 2021. Conversely, in the journal *PLOS One*, the number of publications on PD fell from 145 (2015) to 58 (2021);
3. Identify the most cited journals. Out of 3,292 journals, 2,725 journals had non-zero citations. 661 journals had between 1 and 2 citations of their publications for the entire period, while 59 journals had more than 1,000 citations. The journals with the highest number of citations were "Movement Disorders" and "Parkinsonism & Related Disorders". It is noteworthy that the Top-10 most cited journals include not only those with the largest number of publications, but also journals that do not rank highly by number of publications. For example, "The Lancet Neurology" ranks 8th in terms of citations from 2015 to 2021, while occupying the 83rd place by number of publications on PD;
4. Examine the dynamics of journal citations in publications of other journals, as well as in their own publications.

This approach to the analysis of a scientific field is universal and can be applied to the study of various scientific areas to identify actively developing journal groups and, consequently, research directions.

References

Cowhitt, T., & Cutts, A. (2020). Using network analysis to compare bibliographic database journal coverage. *Journal of Electronic Resources Librarianship, 32*(3), 195–210. https://doi.org/10.1080/1941126X.2020.1790952

Delardas, O., & Giannos, P. (2022). How COVID-19 Affected the journal impact factor of high impact medical journals: Bibliometric analysis. *Journal of Medical Internet Research, 24*(12), Article e43089. https://doi.org/10.2196/43089

Ioannidis, J. (2006). Concentration of the most-cited papers in the scientific literature: Analysis of journal ecosystems. *PLoS ONE, 1*(1), Article e5. https://doi.org/10.1371/journal.pone.0000005

Merigó, J., & Yang, J. (2017). Accounting research: A bibliometric analysis. *Australian Accounting Review, 27*, 71–100. https://doi.org/10.1111/AUAR.12109

Stachura, A., Banaszek, Ł., & Włodarski, P. K. (2023). Should experts guide research merit assessment? Discrepancies between Journal Impact Factor and national expert-based journal ranking list for evaluation of scientific achievement in Poland; a bibliometric analysis. *medRxiv*, https://doi.org/10.1101/2023.10.26.23297621

Zeng, Z., Jia, C., Zhang, W., Zhuang, X., & Li, X. (2024). Assessing journals through a three-dimensional framework based on article citation, author and institution influence. *Journal of Information Science, 50*, 786–807. https://doi.org/10.1177/01655515241238401

Chapter 7
An Analysis of Citations of Publication Authors and Their Affiliations

Citation analysis of an author determines the personal contribution of a scientist to the field of knowledge the author is developing. Citation is an indicator of the authority, usefulness, and effectiveness of researchers, reflecting the acceptance of their work by the scientific community and its influence on subsequent studies. However, some researchers argue that the citation index is a subjective indicator and cannot serve as a measure of the quality of a scientist's work. For example, in (Ding et al., 2020), the effectiveness of the Hirsch index and its versions in assessing individual research activity is critically reviewed. The study emphasizes the need for a multifaceted approach to research evaluation, combining quantitative metrics with qualitative assessments. While traditional bibliometric methods focus on quantitative indicators such as the number of publications and citation indices, network analysis provides a deeper understanding of how knowledge spreads and which institutions play leading roles in shaping scientific directions.

Classic and new centrality indices allowed for the identification of the most important organizations engaged in PD research. Clustering dynamic citation changes using pattern analysis helped identify various citation patterns. Based on pattern analysis, scientists and organizations can draw conclusions about trends in the significance of research centers, while investors can plan capital allocation to specific organizations. Identifying communities allowed for the determination of the main groups of articles closely related to each other and corresponding to specific research areas.

Thus, network analysis methods are essential tools for understanding the structure and dynamics of scientific collaboration between organizations. The use of these methods helps not only to quantitatively assess the contributions of organizations to scientific development but also to identify the qualitative aspects of their influence on the formation of new knowledge. The development of such analytical approaches contributes to increasing the transparency of scientific activity and the efficiency of research processes. Their application can play a key role in shaping science and technology policies, optimizing resource distribution, and creating more sustainable and productive scientific ecosystems.

© The Editor(s) (if applicable) and The Author(s), under exclusive license to Springer Nature Switzerland AG 2026

F. Aleskerov et al., *Bibliometric Analysis by Network Models*, Contemporary Systems Thinking, https://doi.org/10.1007/978-3-032-09171-0_7

Citation network analysis plays a key role in studying scientific communications, allowing the identification of collaboration structures and assessment of the impact of individual publications, authors, and organizations. In (Aria & Cuccurullo, 2017), an R-language package called bibliometrix is presented, designed for comprehensive scientific mapping. This tool provides researchers with capabilities for quantitative analysis of scientific literature, including data import, bibliometric analysis, and network visualization. The research methods include identifying productive authors, determining leading journals, and constructing various bibliometric networks, such as co-citation, collaboration, and word co-occurrence networks. In (Ho & Shekofteh, 2021), a scientometric analysis of highly cited publications on multiple sclerosis theme in the Science Citation Index Expanded (SCIE) is conducted. The Y-index, which evaluates the productivity of researchers, is used to analyze the contribution of authors. In (Yeung et al., 2024), an analysis of the 100 most cited works in the field of digital and mobile health is presented. Research methods included bibliometric analysis of citations of authors, journals, and countries, analysis of interaction networks of authors and their levels of interaction, and identification of the most influential topics and directions.

Modern studies actively use network analysis and machine learning methods for more accurate measurement of scientific impact. Of particular interest is the study of citation networks at the organizational level, as scientific research is often the result of collective work by universities, research centers, and public or private institutions. Analysis of these networks makes it possible to determine the most influential organizations, their interconnections, and to identify structural features of scientific collaboration.

In (Shishehgarkhaneh et al., 2022), a bibliometric and network analysis was conducted to study the application of new technologies in the construction industry. The study focused on analyzing citation networks of organizations, which makes it possible to identify key scientific institutions and organizations that make significant contributions to the development of these technologies. The Web of Science database was used for the analysis, and methods of co-citation and co-occurrence were applied. Using these methods and network visualization, connections between authors, countries, and organizations were analyzed, allowing the identification of major research centers and the establishment of key trends in these areas. The co-citation method was used to identify the most significant publications and studies, as well as to analyze patterns of interaction between different scientific institutions and research groups. The results of the study demonstrated that both leading universities and specialized research centers that actively collaborate with each other are the most influential.

Network analysis plays a key role in bibliometric research, allowing the identification of the structures of scientific communities, the measurement of the influence of individual authors and organizations, as well as the analysis of the dynamics of scientific collaboration. The use of centrality metrics helps to understand which nodes in the network have the greatest influence, how knowledge spreads within the academic community, and what factors may limit access to scientific information. These approaches are applied in the study (Duggan et al., 2024), which analyses the publication activity in the field of anesthesiology. The study uses social network analysis methods to identify patterns of interaction between authors and scientific institutions, as well as to reveal important imbalances in the level of influence of different participants in the network. Centrality analysis was applied to assess the influence of authors and organizations in the network. Several centrality metrics, such as Degree, Eigenvector, and Betweenness, were used to identify the most influential authors and institutions. The results showed that some authors and institutions have significant influence in the network. Using metrics such as the Gini coefficient, the authors also assessed the degree of inequality in the distribution of influence among nodes. The Gini coefficient indicated that there is significant inequality among both authors and institutions, which may affect the distribution of knowledge and opportunities for collaboration. Thus, network analysis in this context makes it possible not only to see the current structure of the scientific network but also to identify directions for improving collaboration, more equitable distribution of information, and inclusion of diverse perspectives in future research. In (Chen et al., 2023), while analyzing the innovation network of the integrated circuit industry in China, the authors using network centrality analysis, revealed that innovation in cities involves close cooperative relationships, including other institutions like universities.

In (Zhou et al., 2023), the evolution of innovative cooperation between industry, universities, and research institutes in the context of low-carbon technology development in China is examined. The research methodology is based on the analysis of patent data and the application of social network analysis methods and community detection algorithms. Three stages of network evolution were identified, demonstrating a gradual decrease in network density and changes in the dynamics of cooperation. The results show that leading enterprises retain key positions in innovation networks, while the participation of universities and research institutes remains limited. The study emphasizes the importance of strengthening the interaction of scientific organizations with industrial enterprises to accelerate low-carbon innovations.

Thus, the use of comprehensive approaches to citation analysis makes it possible to identify key knowledge centers, assess the intensity of interactions, and analyze the evolution of research connections.

This chapter will examine classical and new network centrality indices in their application to the assessment of citations of publication authors and their affiliations. The dynamics of network changes are considered and the stability of the network over time is analyzed.

Citation network of publications authors

The consideration of the graph of publication activity of authors is carried out primarily in order to identify the most significant authors in them. Since we conducted an analysis only of publications on Parkinson's disease published during the period 2015–2021, all the obtained results of the analysis relate only to this topic for the period under consideration.

7.1 Statistical Analysis of Citations of Publications Authors

As noted above (see Chap. 2), an analysis of references to 27,551 authors was conducted.

The distribution of the number of citations of authors for the entire period under consideration (2015–2021) is shown in Fig. 7.1. Most authors (6482) were not cited: they only cited other authors. 5535 authors had from 10 to 50 citations, and only 12 authors were cited more than 500 times.

An analysis was carried out of the citations of authors of downloaded publications in each year under consideration. Table 7.1 shows the number of references in each year for the 12 authors who had more than 500 citations in 2015–2021. Let us recall that we considered only publications from the Microsoft Research scientific publication database. The table provides the author's AuId assigned in the database and the author's affiliation.

As can be seen from Table 7.1, the first line in it is occupied by the author with AuId 2047227543 from the Montreal General Hospital (Department of Neurology, McGill University). He is the first author, in co-authorship with 16 authors, of the article "MDS clinical diagnostic criteria for Parkinson's disease" (2015), which occupies the first place in terms of the number of citations (1563) and the article "MDS research criteria for prodromal Parkinson's disease" (2015), which ranks 4th in terms of citations (473). There were 84 publications in the downloaded database for the period under consideration where this author was a co-author. An analysis of citations of other works by the author showed a spread of references. For example, the article "From Prodromal to Overt Parkinson's Disease: Towards a New Definition

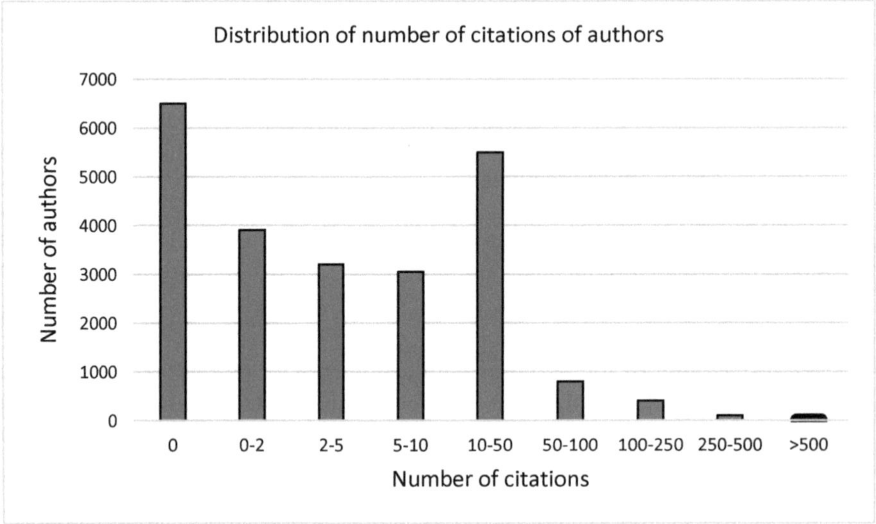

Fig. 7.1 Distribution of the number of citations of authors

Table 7.1 Number of citations of the Top-12 authors in each year

Author Id	Affiliation	Number of citations by years							Total citations
		2015	2016	2017	2018	2019	2020	2021	
2047227543	Montreal General Hospital	15	79	247	290	423	632	730	2416
2064657699	University College London	4	19	40	106	147	171	243	730
1761635969	National Institutes of Health	2	23	30	36	112	229	265	697
2303827053	University of Rochester	0	2	11	15	55	269	344	696
78852038	University of Cincinnati	2	7	44	59	91	210	185	598
2600996391	Biochemistry Section. Surgical Neurology Branch. National Institute of Neurological Disorders and Stroke	19	54	86	84	83	108	127	561
2570671491	Max Planck Society	0	25	47	91	117	142	129	551
2100557140	Veterans Health Administration	12	55	87	88	104	98	106	550
2153866458	University of Padua	5	34	40	69	104	135	136	523
2175448217	German Center for Neurodegenerative Diseases	3	33	108	63	86	110	110	513
2112352720	California Institute of Technology	0	1	46	79	103	125	155	509
2265103992	Katholieke Universiteit Leuven	10	54	63	87	118	87	89	508

in the Year 2040," published in the "Journal of Parkinson's Disease" in 2018, had 5 citations in the downloaded publications, and the article "Research criteria for the diagnosis of prodromal dementia with Lewy bodies" ("Neurology" journal, 2020) had 22 citations. Let us present a few more examples of citations of authors from the list presented in Table 7.1.

The author with AuId 2064657699 from University College London, who took the second line in the list (Table 7.1), had 23 publications in the database and 730 citations to them. The most cited article in the database was "Non-motor features of Parkinson disease," published in the journal "Nature Reviews Neuroscience" in 2017—459 citations. Another article by the author, "Glucocerebrosidase mutations and synucleinopathies: Toward a model of precision medicine," also published in 2017 in the journal "Movement Disorders," was cited 34 times.

The third line in Table 7.1 was taken by the author with AuId 1761635969 from the National Institutes of Health, who had 91 articles in the database and 697 citations. More than half of this number of citations was received by the article "Identification of novel risk loci, causal insights, and heritable risk for Parkinson's disease: a meta-analysis of genome-wide association studies" (journal "The Lancet Neurology," 2018)—373 citations. At the same time, the article "The Parkinson's Phenome: Traits Associated with Parkinson's Disease in a Large and Deeply Phenotyped Cohort" ("Biorxiv," 2018) had 2 citations, and the 2019 article "The Parkinson's Disease Mendelian Randomization Research Portal" ("Movement Disorders") was cited 18 times.

Let us consider two more examples characterizing the number of citations—for the author with the least number of publications and the author with the largest number of publications in the Top-12.

The author with AuId 2303827053 from the University of Rochester had 22 publications and 696 citations for 2015–2021. The article "Global, regional, and national burden of Parkinson's disease, 1990–2016: a systematic analysis for the Global Burden of Disease Study 2016" ("The Lancet Neurology," 2018) had 459 citations. In the same year 2018, the article "Using Smartphones and Machine Learning to Quantify Parkinson Disease Severity: The Mobile Parkinson Disease Score," published in the journal "JAMA Neurology," had 69 citations. The article "Patient Views on Telemedicine for Parkinson Disease," published in 2019 in the "Journal of Parkinson's Disease," brought the author another 17 citations.

The author with AuId 2175448217 from the German Center for Neurodegenerative Diseases had 181 publications and 513 citations to them. The article "MDS research criteria for prodromal Parkinson's disease" ("Movement Disorders," 2015), where the author is the first in the list of 20 authors, was cited 473 times. The article "Movement disorder society criteria for clinically established early Parkinson's disease" in the journal "Movement Disorders," 2018, had 31 citations, and the publication "Prodromal Parkinson disease subtypes – key to understanding heterogeneity," published in April 2021 in the journal "Nature Reviews Neurology," had 4 citations at the time of accessing the scientific publication database.

The conducted citation analysis of authors makes it possible to highlight publications that received the highest or lowest peer evaluation.

7.2 Citation Graph of Publication Authors

For each article, the first author from the list was selected, since, as a rule, in medical articles the first listed author is the lead author or the project supervisor. This is the vertex of the graph. An edge of the graph corresponds to a citation in an article published by author A of a publication by author B. The weight of the edge is the

Table 7.2 Example of a citation network of publication authors

AuId1	AuId2	Weight	Year
1093168	2567903964	1	2018
1873123	1873123	1	2018
1873123	1950300814	1	2017
275834728	2086437927	2	2017
2068236351	2123401699	3	2018

number of citations in a given year. Self-citation is possible in such a network, since an author can cite their own publications.

Network attributes:

- AuId1—ID of the citing author A
- AuId2—ID of the cited author B
- Y—year of publication of the article
- weight—edge weight, that is, the number of citations by author A of author B in year Y

A sample network is shown in Table 7.2.

It should be noted that in the constructed graph, the maximum edge weight in the network is 29, i.e., one of the authors received 29 citations in a certain year. The network includes 27 authors who cited one specific author more than 10 times per year. Only 4 of them cited someone other than themselves. In almost all such cases, these are co-authors.

7.2.1 Statistical Analysis of the Author Citation Network

Table 7.3 presents some indicators of the author citation network. It can be observed that compared to a similar network for journals, this network is significantly less dense, contains an order of magnitude more connected components, and an order of magnitude lower average number of citations.

The constructed network includes 82 connected components. The largest one contains 27,402 authors. The network also has 58 components with author pairs, 5 components with author triplets, and 18 components consisting of a single author who cites only themselves. All author triplets include authors from different organizations, but their works are about similar specialized topics. A similar pattern can be seen in components of size 2. Mainly, these authors specialize narrowly in specific issues related to PD.

For the analysis of centrality in the author citation network, the same quota values were chosen as for journals, and a quota of 0.1% of the total number of journal citations was added.

Table 7.3 Indicators of the citation network of publication authors

Network metrics	Values
Nodes	27,551
Edges	271,623
Number of connected components	82
Largest connected component	27,402
Density	0.0004
Minimum citations	1
Maximum citations	2416
Average citations	14.75

7.3 An Analysis of Citation Network Indices for Publication Authors

Various citation indices allow for a comprehensive assessment of the significance of authors within the scientific publication network. They make it possible to assess the popularity of authors based on the number of citations and to identify the importance of an author based on who is citing them.

Four classical indices were considered: In-degree, Betweenness, Eigenvector, and Pagerank, as well as two new centrality indices: Bundle (BI) and Pivotal (PI).

7.3.1 Classical Centrality Indices

Table 7.4 shows the ten authors with the highest In-degree index values. Authors are ranked by the In-degree index. The table presents the authors' affiliations and their ranks by the Betweenness, Eigenvector, and Pagerank indices.

As can be seen from Table 7.4, the top position is occupied by the author with AuId 2047227543 from Montreal General Hospital (Department of Neurology, McGill University). He is the first author, in collaboration with 16 co-authors, of the article "MDS clinical diagnostic criteria for Parkinson's disease" (2015), which ranks first in terms of citation count (1563), and the article "MDS research criteria for prodromal Parkinson's disease", which ranks fourth in terms of citation count (471). The large number of citations to these articles secured the author the highest rank. This rank is maintained by the author in all subsequent years after the publication of the articles (2015), as are the ranks of these articles. Figure 7.2 shows the annual dynamics of the In-degree index for the top five authors from Table 7.4.

It should be noted that for the first author with AuId 2047227543, the value of the index throughout the entire period was significantly higher than the others. The Top-10 authors ranked by Eigenvector index value are presented in Table 7.5. The value of this index determines the influence of references. The more "influential" publications cite this author, the more "influential" the author is.

Table 7.4 Top-10 authors by number of citations (by In-degree index)

№	Author Id	Affiliation	In-degree values	Rank		
				Betweenness	Eigenvector	Pagerank
1	2047227543	Montreal General Hospital	0.0078	2	1	1
2	2064657699	University College London	0.0023	116	31	5
3	2303827053	University of Rochester	0.0022	25	78	3
4	1761635969	National Institutes of Health	0.0022	201	13	10
5	78852038	University of Cincinnati	0.0019	6	18	24
6	2600996391	National Institute of Neurological Disorders and Stroke (NINDS)	0.0018	10,751	411	2
7	2570671491	Max Planck Society	0.0018	320	224	42
8	2100557140	Veterans Health Administration	0.0018	10	11	4
9	2153866458	University of Padua	0.0017	4	6	27
10	2175448217	German Center for Neurodegenerative Diseases	0.0017	7	2	20

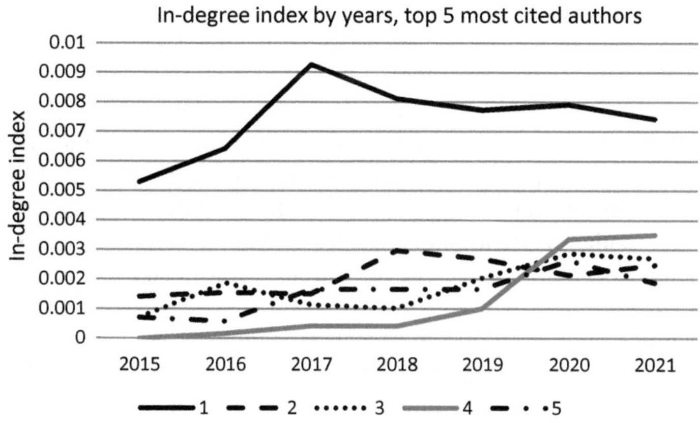

Fig. 7.2 Dynamics of changes in the In-degree index of the Top-5 authors by year

Table 7.5 Top-10 authors by Eigenvector index

№	Author Id	Affiliation	Eigenvector values	Rank		
				Betweenness	In-degree	Pagerank
1	2047227543	Montreal General Hospital	0.6725	2	1	1
2	2175448217	German Center for Neurodegenerative Diseases	0.2737	7	10	20
3	2099613546	UCL Institute of Neurology	0.1866	33	18	13
4	2599128148	Karolinska Institutet	0.1825	279	37	75
5	2732209544	UCL Institute of Neurology & University of Innsbruck	0.1801	73	43	70
6	2153866458	University of Padua	0.1613	4	9	27
7	2026404387	Tel Aviv Sourasky Medical Center	0.1306	70	17	24
8	2049643778	King's College London	0.1282	222	60	201
9	3183379636	McGill University	0.1134	37	40	122
10	1929534198	Mayo Clinic	0.1077	390	165	259

This list of authors differs from the list presented in Table 7.4. As can be seen from the table, only three authors who entered the Top-10 by In-degree index are included in this list. Author No. 1 (AuId 2047227543) retains first place also by Eigenvector index. Author No. 2 (AuId 2175448217) had the 10-th rank by In-degree index, and Author No. 6 (AuId 2153866458) occupied the 9-th position by In-degree index (Table 7.4). The table features the same author (AuId 2599128148 and AuId 3183379636) with different affiliations and different index values (No. 4 and 9). Until 2016, this author's articles indicated one affiliation—Karolinska Institutet, and in subsequent years the author lists two affiliations—Karolinska Institutet and McGill University.

The dynamics of changes in the Eigenvector index from 2015 to 2021 for the Top-5 authors are shown in Fig. 7.3. A rise in the index for these authors is noted in 2016–2017 and in 2019. The minimum index value among the authors was in 2018. Comparing this with the publication dynamics (see Chap. 3, Fig. 3.1), we also note an increase in the number of publications from 2015 to 2017 and from 2019 to 2021. However, the index of authors that rose in 2019 declined during the period 2020–2021.

Table 7.6 presents the Top-10 authors by the PageRank index. The results for this index most closely resemble the Top-10 by In-degree index (see Table 7.4). Six

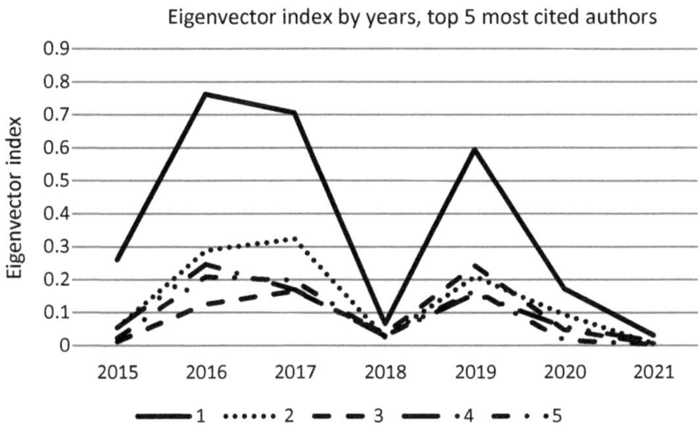

Fig. 7.3 Dynamics of changes in the Eigenvector index of the Top-5 authors by year

authors from the Top-10 list by In-degree index entered the Top-10 by PageRank index.

It can be noted that the PageRank value for author No. 1 (AuId 2047227543) is 0.0062, while the values for other authors range from 0.001 to 0.002. Figure 7.4 shows the dynamics of changes in the PageRank index for the first five authors from Table 7.6.

Interestingly, author No. 2 (AuId 2600996391) had the maximum index value in 2015, i.e., was the most influential author, but over 7 years the index value decreased annually. Of all the authors in the Top-5, only author No. 3 (AuId 2303827053) had a positive trend in the PageRank index value.

Table 7.7 shows the Top-10 authors by Betweenness index. This is the only index where the highest value does not belong to author No. 2 (AuId 2047227543). Author No. 1 (AuId 260914988), who had not previously entered the Top-10 authors by other classical indices, leads this list.

Let us consider how the Betweenness index changed by year for the first five authors presented in Table 7.7. Figure 7.5 presents the dynamics of changes in the Betweenness index for the Top-5 authors, and Table 7.8 shows the yearly Betweenness index values for these authors.

From Table 7.8 we can observe there are years when the Betweenness index value for some authors is zero. This is explained by the fact that in that year, the respective author published articles without references to other authors in the studied network, and also was not cited by other authors from this network. Attention should be paid to the dynamics of changes in the Betweenness index (Fig. 7.5) for author No. 1 (AuId 2609614988) and author No. 2 (AuId 2047227543). In 2019, both authors were important sources of information, acting as intermediaries between other authors in the network, receiving and giving references.

Table 7.6 Top-10 authors by citation count according to PageRank index

№	Author Id	Affiliation	PageRank values	Rank		
				Betweenness	In-degree	Eigenvector
1	2047227543	Montreal General Hospital	0.0062	2	1	1
2	2600996391	National Institute of Neurological Disorders and Stroke (NINDS)	0.0020	10,751	6	411
3	2303827053	University of Rochester	0.0015	25	4	78
4	2100557140	Veterans Health Administration	0.0014	10	8	11
5	2064657699	University College London	0.0013	116	2	31
6	1761635969	National Institutes of Health	0.0013	201	3	13
7	2195539447	Haukeland University Hospital	0.0012	451	16	1080
8	1991986455	Harvard University	0.0012	167	19	276
9	2243759912	CEU San Pablo University	0.0012	8	30	428
10	2265103992	Katholieke Universiteit Leuven	0.0011	51	12	52

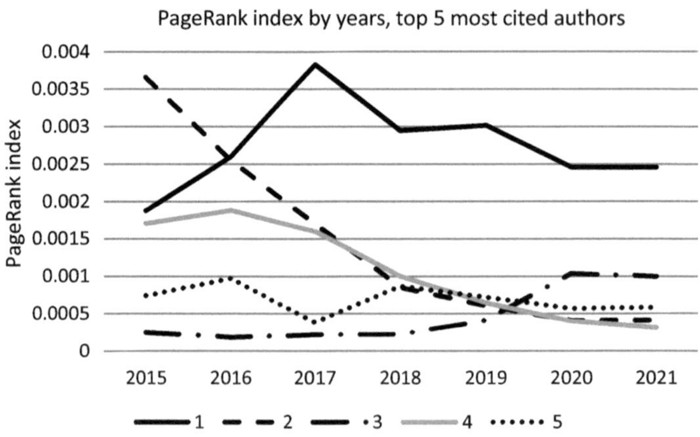

Fig. 7.4 Dynamics of changes in the PageRank index for the Top-5 authors

Table 7.7 Top-10 authors by citation count according to Betweenness index

№	Author Id	Affiliation	Betweenness values	Rank		
				Pagerank	In-degree	Eigenvector
1	2609614988	Institute of Clinical Neurobiology. Vienna. Austria	0.029	107	61	25
2	2047227543	Montreal General Hospital	0.024	1	1	1
3	2127596906	Toronto Western Hospital	0.013	18	15	14
4	2153866458	University of Padua	0.01188	27	9	6
5	2165933549	Veterans Health Administration	0.01141	793	376	1029
6	78852038	University of Cincinnati	0.0111	21	5	18
7	2175448217	German Center for Neurodegenerative Diseases	0.0107	20	10	2
8	2243759912	CEU San Pablo University	0.00911	9	30	428
9	2170598338	Oregon Health & Science University	0.00896	72	69	280
10	2100557140	Veterans Health Administration	0.00895	4	8	11

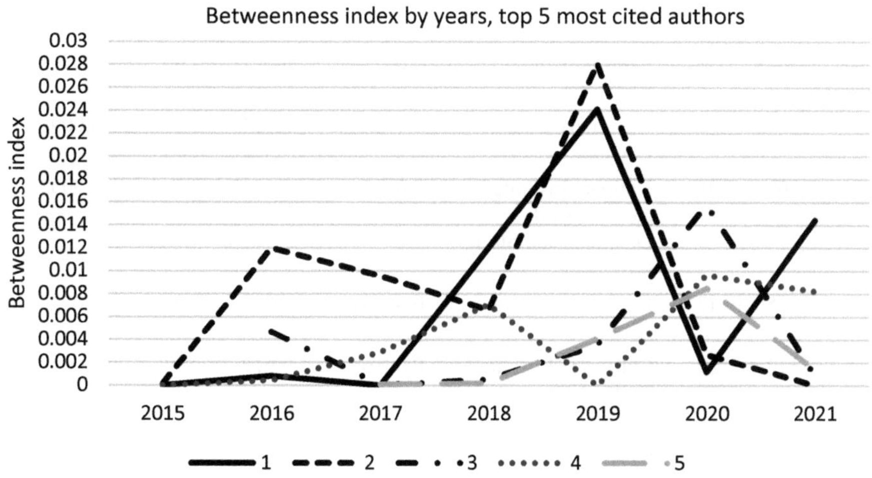

Fig. 7.5 Dynamics of changes in the Betweenness index for the Top-5 authors

Table 7.8 Values of the Betweenness index for the Top-5 authors by year

№	AuId	Betwenness values						
		2015	2016	2017	2018	2019	2020	2021
1	2609614988	0	0.000819	0	0.0120	0.0241	0.0012	0.0144
2	2047227543	0.000127	0.01198	0.0095	0.0066	0.0281	0.0027	0
3	2127596906	0	0.00469	0.000029	0.00052	0.0035	0.0158	0.0009
4	2153866458	0.00000065	0.00045	0.00289	0.00707	0	0.0096	0.0082
5	2165933549	0	0	0.000087	0.00018	0.0041	0.0085	0.0014

7.3.2 New Centrality Indices

Table 7.9 presents the Top-10 authors by BI, PI, and TI indices with a quota of 0.1%, ordered by the TI index. As in the case of journals, the results are almost identical to the Top-10 by the In-degree index or the list of most cited authors. It is interesting that the PI index of the first author (AuId 2047227543) is nearly equal to 1, while the other authors have values close to zero. This means that the first author has the highest number of pivotal nodes, an order of magnitude greater than the others.

Table 7.9 Top-10 authors by citation count by TI index $q = 0.1\%$

№	Author Id	Affiliation	In-degree values	$q = 0.1\%$		
				BI	PI	TI
1	2047227543	Montreal General Hospital	0.0071	0.577	0.99	0.528
2	2303827053	University of Rochester	0.0023	0.0251	0	0.0091
3	2064657699	University College London	0.0024	0.0245	0	0.009
4	2600996391	National Institute of Neurological Disorders and Stroke (NINDS)	0.0018	0.0142	0	0.0053
5	2195539447	Haukeland University Hospital	0.00164	0.0122	0	0.0046
6	2112352720	California Institute of Technology	0.00168	0.012	0	0.00457
7	78852038	University of Cincinnati	0.0018	0.011	0	0.0043
8	1761635969	National Institutes of Health	0.00197	0.0104	0	0.0041
9	1991986455	Harvard University	0.0016	0.0105	0	0.004
10	2100557140	Veterans Health Administration	0.0017	0.0096	0	0.0038

Table 7.10 shows a similar result for a quota of 1%. It can be noted that this table has no matches with the tables for classical indices (Tables 7.4–7.7) and TI with a quota of 0.1% (Table 7.9). The values of the In-degree index are an order of magnitude lower than in Table 7.4. Unlike Table 7.9, the values are within almost the same range, and there is no author that stands out significantly from the others.

Table 7.11 presents the Top-10 authors by TI with the highest quota of 10%. In this list, the most cited author (No. 1 AuId 2047227543) again has the maximum value. At the 5% quota, he is also in first place. However, he has zero values for the new indices, since he has such a large number of citations that a 10% quota becomes too large. Other authors in the table have identical index values. This can be explained by the fact that in the author citation network, edge weights are lower than in the journal citation network, so for a large quota the number of critical sets and pivotal nodes is much smaller. Table 7.11 includes three authors from German universities and two from Chinese universities—they cite each other more intensively than others.

The dynamics of changes over the years in BI and PI indices for authors are presented using the example of quota $q = 1\%$. Figure 7.6 shows the changes in the BI index values from 2015 to 2021 for the Top-5 authors from Table 7.10.

Let us consider the index of author AuId 2141423704, who occupies the 4th position in Table 7.9. From 2015 to 2018, the author's index was declining. This may indicate a decrease in the collective influence of the group of authors with whom he

Table 7.10 Top-10 authors by citation count with quota $q = 1\%$

№	Author Id	Affiliation	In-degree values	$q = 1\%$		
				BI	PI	TI
1	2303925656	Chinese Academy of Sciences	0.00093	0.0129	0.0306	0.0150
2	3184490928	Broomfield Hospital	0.00094	0.0129	0.03	0.0146
3	2731932222	Northwestern University	0.00097	0.0124	0.0267	0.0134
4	2141423704	Van Andel Institute	0.00095	0.0117	0.0252	0.0126
5	2618592063	University College London	0.00089	0.0102	0.0224	0.0112
6	1997750176	Fondazione I.R.C.C.S. Istituto Neurologico Carlo Besta	0.0009	0.01	0.0214	0.0108
7	2170598338	Oregon Health & Science University	0.00087	0.009	0.02	0.010
8	2135251184	Rush University Medical Center	0.00087	0.0089	0.019	0.0096
9	328685278	Northwestern University	0.00087	0.0087	0.0183	0.0093
10	2009196669	Saarland University	0.00081	0.008	0.0185	0.0091

Table 7.11 Top-10 authors by citation count by TI index with quota $q = 10\%$

№	Author Id	Affiliation	In-degree values	$q = 10\%$		
				BI	PI	TI
1	2047227543	Montreal General Hospital	0.0071	0	0	0.0024
2	1887309180	Ohio State University	0.0001	0.0009	0.0015	0.0009
3	2575752043	Chinese Academy of Sciences	0.0001	0.0009	0.0015	0.0009
4	2096830223	Okayama University	0.0001	0.0009	0.0015	0.0009
5	2557304392	Medical University of Graz	0.0001	0.0009	0.0015	0.0009
6	2894855075	University of Göttingen	0.0001	0.0009	0.0015	0.0009
7	2898120263	National Taiwan University	0.0001	0.0009	0.0015	0.0009
8	2279503223	New Generation University College	0.0001	0.0009	0.0015	0.0009
9	2280740378	Capital Medical University	0.0001	0.0009	0.0015	0.0009
10	2467486701	Saarland University	0.0001	0.0009	0.0015	0.0009

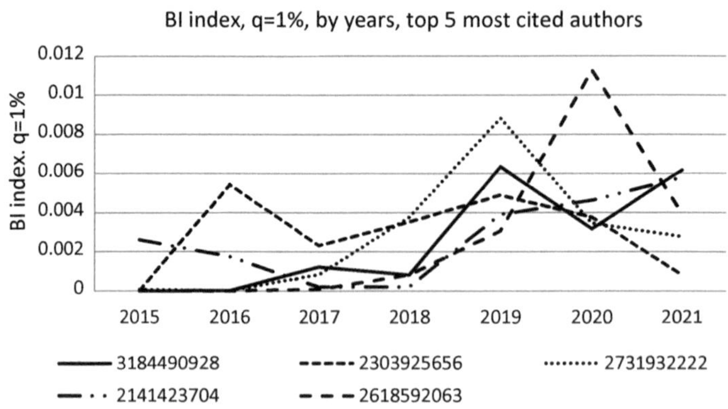

Fig. 7.6 Dynamics of changes in the BI index over the years

actively collaborated, or a reduction in his role in key citation groups. However, since 2019, the index has shown steady growth. This may indicate the strengthening of this author's role in the research network, possibly through the resumption of active work or a return to significant scientific collaborations, which again led to his influence on a broader circle of researchers.

Author No. 3 (AuId 2731932222) in Table 7.9 had an index value close to 0 before 2017, which indicates low collective influence during that period. However, from 2017 to 2019 the index value increased significantly. This may be explained by the fact that during this period the author became part of more influential research groups or his works began to gain collective recognition from more significant authors. Since 2019, the index has begun to decline again, which may be due to a weakening of his role in collaborations or a decrease in citation of his works by other key authors.

The dynamics of changes in PI index values for authors are shown in Fig. 7.7. In contrast to the BI index for author AuId 2141423704 (No. 4 in Table 7.10), the PI index, having dropped from 2015 to 2017, did not recover in subsequent years, remaining close to zero. This may indicate that although the author again became an important participant in collaborations (BI growth), his role as an intermediary connecting different clusters or groups of researchers has significantly decreased. The importance of this author in maintaining links between different research clusters has likely weakened, and he no longer plays a key role in integrating different directions in the network. This may indicate that his research is becoming more localized or specialized, influencing only a limited circle of authors rather than the entire network.

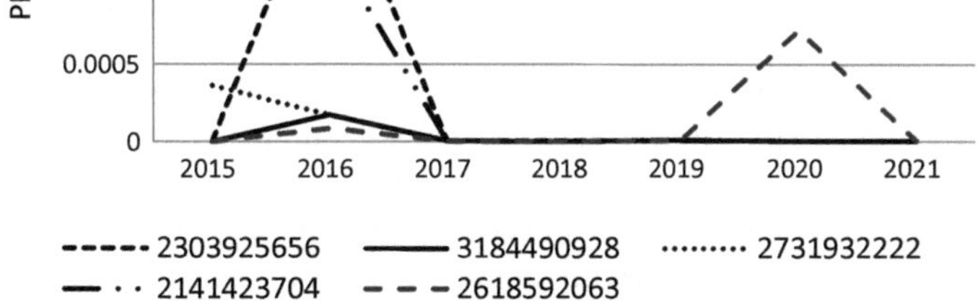

Fig. 7.7 Dynamics of changes in the PI index over the years

7.4 Main Conclusions from the Analysis of Author Citations

Analyzing the affiliation of authors to various organizations from the tables above, it can be concluded that there is no clear correlation between the most influential authors and their affiliations. The highest centrality index values are found among the most renowned universities in the world, while authors from these institutions do not appear in the lists of authors with the highest centrality indices.

Based on this analysis, it can be concluded that different centrality indices allow for different assessments of the significance of authors in the scientific publication network. The In-degree index reflects the direct popularity of authors based on the number of citations, whereas PageRank and Eigenvector indicate more complex aspects of their influence by taking into account the importance of those who cite them. Betweenness, in turn, highlights authors who play a key role in connecting different research groups, making them important figures in the dissemination of knowledge between clusters. The use of the BI and PI indices with different quotas allows for a deeper investigation of the level of influence of authors at various scales—from small but highly cited groups to broader and more extensive segments. Thus, the use of these indices in combination makes it possible to obtain a more complete and multifaceted picture of researchers' influence on the scientific community and the development of respective fields of knowledge.

7.5 An Affiliation Citation Graph

The effectiveness of a scientific organization's activity is determined by the success of its employees, including the number of their publications in high level journals with large number of citations, where the given organization is mentioned. It is important to highlight scientific centers and organizations that are leaders in particular scientific fields.

To construct the affiliation graph, an analysis of the articles was conducted. Among the articles included in the citation network, not all contained the affiliation identifier of the authors—some lacked the name of the affiliation, others contained errors or incorrect formats (e.g., only the city or country), which made it impossible to determine the affiliation. 5631 articles without identifiers were removed from the network. When constructing the citation network of affiliations for author teams, the affiliation of the first author listed among all authors was taken into account. As a rule, the first author leads the research group and often this organization funds the research. As a result, the affiliation network comprised 3029 nodes and 170,340 edges. This number of edges included parallel edges when one organization cited another in different years. When grouping these edges by year, their number decreased to 128,602.

The network has the following structure—an edge between affiliation AfId1 and AfId2 indicates that a paper with affiliation AfId1 cites a paper with affiliation AfId2

Fig. 7.8 Fragment of the affiliation citation network

AfId1	AfId2	Y	Weight
4605	138006243	2017	1
9507	9507	2021	2
9507	35028602	2021	1
9507	40120149	2021	1
9507	177639307	2021	1

(Fig. 7.8). Here, Id is a numerical identifier assigned by the Microsoft Academic scientific publication search system (Sinha et al., 2015; Wang, 2019). The weight of the edge is the number of citations in a given year.

Network attributes:

AfId1—Id of the citing organization
AfId2—Id of the cited organization
Y—year of citation of papers with affiliation AfId2 by papers with affiliation AfId1
Weight—number of references from AfId1 to AfId2 in year Y, i.e., the weight of the edge

The constructed network consists of five connected components. The largest connected component contained 3025 affiliations. The remaining four contain one isolated affiliation with self-citation, without other edges

- Universidad del Tolima, Tolima, Colombia
- Northern Kentucky University, Kentucky, U.S.
- Chiba Institute of Science, Chiba, Japan
- Savitribai Phule Pune University, Pune, Maharashtra, India

7.5.1 Statistical Analysis of the Affiliation Citation Network

The affiliation citation network is smaller than the article citation network but denser, since affiliations can cite each other more than once. The network characteristics are presented in Table 7.12.

The analysis of the affiliation citation network for 2015–2021 showed that 439 affiliations were not cited, 297 had one citation, 113 affiliations had more than 500 citations, and 39 of them were cited more than 1000 times. Three research centers had more than 3000 citations. Table 7.13 presents the Top-20 affiliations with the highest number of citations.

The list mainly includes universities in the United Kingdom, USA, Belgium, and Sweden. They are among the largest and oldest in the world, with many Nobel laureates working in them, and they conduct research in various scientific fields. The National Institutes of Health (No. 1) consists of 27 national health research institutes and centers in the USA and is the main center in the USA responsible for medical

Table 7.12 The characteristics of the affiliation citation network

Name of the indicator	Value
Number of nodes	3029
Number of edges	128,602
Number of connected components	5
Density	0.014
Minimum number of citations	0
Maximum number of citations	3757
Average number of citations	79.6

Table 7.13 Top-20 affiliations with the highest number of citations

Rank	Affiliation	Number of citations 2015–2021rr
1	National Institutes of Health	3757
2	UCL Institute of Neurology	3233
3	University of Cambridge	3165
4	University of Oxford	2933
5	University College London	2882
6	Northwestern University	2759
7	Harvard University	2413
8	University of Pennsylvania	2046
9	Katholieke Universiteit Leuven	1927
10	Karolinska Institutet	1893
11	King's College London	1885
12	Capital Medical University	1878
13	Mayo Clinic	1863
14	Radboud University Nijmegen	1799
15	Lund University	1793
16	Newcastle University	1702
17	University of Alabama at Birmingham	1694
18	Shanghai Jiao Tong University	1677
19	University of California, San Francisco	1634
20	University of Sydney	1592

research. UCL Institute of Neurology (No. 2) is an institute of the Faculty of Brain Sciences at University College London. Karolinska Institutet in Sweden (No. 10) is one of the largest medical universities in Europe. The list includes one of the oldest universities in China—Shanghai Jiao Tong University (No. 18), ranked 5th among Chinese universities. There is one of the oldest universities in Australia—the University of Sydney (No. 20).

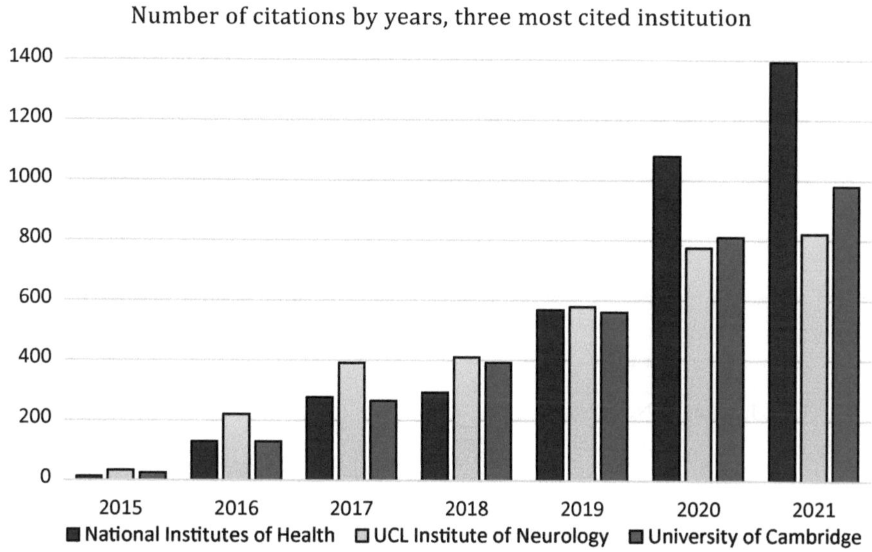

Fig. 7.9 Dynamics of the number of citations in 2015–2021 for the Top-3 affiliations

Dynamics of the change in the number of citations for the Top-3 affiliations is presented in Fig. 7.9. We see that the number of citations in the period 2015–2021 is increasing for the affiliations presented. At the same time, citations to the National Institutes of Health are growing faster. And while in 2015 the Institute ranked 12th in terms of citations, in 2020 it rose to 1st place and retained it in 2021. The UCL Institute of Neurology of University College London ranked first in the number of citations from 2016 to 2018; however, in 2020 it dropped to 5th position, and in 2021 to sixth place in terms of citations. Over the entire period from 2015 to 2021, the UCL Institute of Neurology ranked second in citations overall.

7.5.2 *An Analysis of the Publication Activity of Affiliations for 2015–2021*

We analyzed in which journals and how many times publications related to Parkinson's disease appeared, indicating the respective affiliation. It should be recalled that for research teams only the affiliation of the first author listed among all authors was taken into account. Table 7.14 presents the Top-10 affiliations by number of citations, indicating the number of journals and the number of publications in each of them where this affiliation is presented.

As can be seen from the table, there are journals where research centers publish more often than others. We analyzed the number of publications from these research

Table 7.14 Publication activity of the Top-10 research centers over 7 years (2015–2021)

Rank	Institutions	Number of journals	Number of publications in each journal
1	National Institutes of Health	93	1
		28	2
		11	3
		2	4
		4	5
		1	6
		2	8
		1	12,13,16,19
2	UCL Institute of Neurology	59	1
		15	2
		11	3
		3	4
		2	9
		1	7,10,12,21
3	University of Cambridge	91	1
		21	2
		7	3
		7	4
		2	5
		3	7, 8
		1	6,9,11,14
4	University of Oxford	69	1
		13	2
		7	3,4
		4	5
		2	10
		1	19,20
5	University College London	62	1
		20	2
		5	3
		4	4
		2	5
		1	6,7,10,15,20
6	Northwestern University	64	1
		16	2
		3	3

(continued)

Table 7.14 (continued)

Rank	Institutions	Number of journals	Number of publications in each journal
		3	4
		2	8
		1	5,6
7	Harvard University	70	1
		21	2
		3	3
		2	6
		2	8
		1	4,5,10
8	University of Pennsylvania	87	1
		14	2
		6	3
		2	4,5
		2	6
		1	12,14
9	Katholieke Universiteit Leuven	39	1
		16	2
		7	3
		3	5
		1	4,6,7
10	Karolinska Institutet	78	1
		17	2
		8	3
		6	4
		2	6,10
		1	7,9

centers in high-ranking journals. Table 7.15 presents a list of journals where five or more articles from the Top-10 research centers were published.

In Chap. 5, which deals with the analysis of journal citations, we identified three journals with the highest number of publications on Parkinson's disease research during 2015–2021. These journals are "Movement Disorders" (1201 articles), "Parkinsonism & Related Disorders" (1260 publications), and "bioRxiv" (795 publications). As can be seen from Table 7.15, these journals are the leaders in the number of publications for most of the Top-10 research centers. An analysis of the data in Table 7.15 based on the thematic focus of the journals allows us to draw conclusions about the research orientation of each center. For example, if the University of Cambridge publishes widely in journals with various thematic directions from

Table 7.15 Number of publications in high-ranking journals by the Top-10 research centers (2015–2021)

Rank	Institutions	Highly cited journals	Number of articels
1	National Institutes of Health	Movement Disorders	16
		bioRxiv	13
		medRxiv	12
		Parkinsonism & Related Disorders	9
		Neurobiology of Aging	8
		Neurobiology of Disease	8
		Human Molecular Genetics	6
2	UCL Institute of Neurology	Movement Disorders	21
		Brain	12
		Journal of Parkinson's Disease	10
		Journal of Neurology, Neurosurgery and Psychiatry	9
		Parkinsonism & Related Disorders	9
		Neuropathology and Applied Neurobiology	7
3	University of Cambridge	bioRxiv	14
		Movement Disorders	11
		Brain	9
		Proceedings of the National Academy of Sciences of the United States of America	8
		Scientific Reports	8
		Journal of Parkinson's Disease	8
4	University of Oxford	bioRxiv	20
		Brain	18
		Journal of Neuroscience	10
		Movement Disorders	10
5	University College London	Movement Disorders	20
		bioRxiv	15
		JAMA Neurology	10
		Brain	7
		medRxiv	6
		Journal of Parkinson's Disease	5

(continued)

Table 7.15 (continued)

Rank	Institutions	Highly cited journals	Number of articels
6	Northwestern University	Movement Disorders	8
		bioRxiv	8
		Neuron	6
		Journal of Parkinson's Disease	5
7	Harvard University	Movement Disorders	10
		Neurology	8
		Proceedings of the National Academy of Sciences of the United States of America	8
		JAMA Neurology	6
		Journal of Parkinson's Disease	6
		Parkinsonism & Related Disorders	5
8	University of Pennsylvania	Parkinsonism & Related Disorders	14
		Movement Disorders	12
		Movement Disorders Clinical Practice	6
		Journal of Parkinson's Disease	6
		Neurology	5
		bioRxiv	5
9	Katholieke Universiteit Leuven	PLOS One	7
		Neurorehabilitation and Neural Repair	6
		Frontiers in Neuroscience	5
		Parkinsonism & Related Disorders	5
		Movement Disorders	5
		Gait & Posture	4
10	Karolinska Institutet	Movement Disorders	10
		bioRxiv	10
		Journal of Parkinson's Disease	9
		npj Parkinson's Disease	7
		Brain	6
		Parkinson's Disease	6

'Movement Disorders' to "Scientific Reports," the University of Oxford has many publications in the journals "Brain" and "Journal of Neuroscience," covering research in all aspects of neurology and neuroscience.

7.6 An Analysis of Network Citation Indices of Affiliations

7.6.1 Classical Centrality Indices

The In-degree index value corresponds to the normalized number of citations or incoming edges. Table 7.16 presents the Top-10 cited affiliations ranked by the In-degree index value. For comparison, the table provides rankings of the Betweenness, Eigenvector, and PageRank indices for these affiliations.

Seven research centers are included in the Top-10 for all analyzed centrality indices. A comparative analysis was carried out between the affiliations of the Top-10 publications and the affiliations included in the top by all indices. Table 7.17 presents rankings by In-degree for the Top-10 most cited publications and their affiliations.

The analysis showed that only four research centers published articles in the Top-10 from 2016 to 2021. Thus, high In-degree values for affiliations can be achieved by producing a large number of articles with not necessarily the highest citation ratings.

The dynamics of the In-degree index values for 2015–2021 are illustrated (Fig. 7.10) using the example of the last three research centers from the Top-10

Table 7.16 In-degree index value for the Top-10 affiliations

№	Institutions	In-degree values	Rank		
			Betweenness	Eigenvector	Pagerank
1	National Institutes of Health	0.0156	5	1	1
2	UCL Institute of Neurology	0.0134	3	4	2
3	University of Cambridge	0.0131	1	3	4
4	University of Oxford	0.0122	2	2	3
5	University College London	0.0120	6	5	5
6	Northwestern University	0.0114	14	6	6
7	Harvard University	0.0100	8	8	7
8	University of Pennsylvania	0.0085	24	7	8
9	Katholieke Universiteit Leuven	0.0080	26	10	12
10	Karolinska Institutet	0.0079	10	9	9

Table 7.17 Affiliations of the Top-10 publications

In-degree rank of articles	Title of the article	Year of publ	Affiliation	In-degree rank of affiliation
1	MDS clinical diagnostic criteria for Parkinson's disease	2015	Montreal General Hospital	22
2	Gut Microbiota Regulate Motor Deficits and Neuroinflammation in a Model of Parkinson's Disease	2016	California Institute of Technology	111
3	Epidemiology of Parkinson's disease	2017	Harvard University	7
4	MDS research criteria for prodromal Parkinson's disease	2015	German Center for Neurodegenerative Diseases	37
5	The epidemiology of Parkinson's disease: risk factors and prevention Harvard University	2015	Harvard University	7
6	Non-motor features of Parkinson disease	2017	University College London	5
7	Global, regional, and national burden of Parkinson's disease, 1990–2016: a systematic analysis for the Global Burden of Disease Study 2016	2018	University of Rochester	80
8	The Roles of PINK1, Parkin and Mitochondrial Fidelity in Parkinson's Disease	2015	Biochemistry Section, Surgical Neurology Branch, National Institute of Neurological Disorders and Stroke (NINDS),	1
9	α-Synuclein strains cause distinct synucleinopathies after local and systemic administration	2015	KU Leuven, Laboratory for Neurobiology and Gene Therapy, Department of Neurosciences	9
10	Gut microbiota are related to Parkinson's disease and clinical phenotype	2015	Department of Neurology Helsinki University Central Hospitaland Department of Neurological Sciences University of Helsinki	65

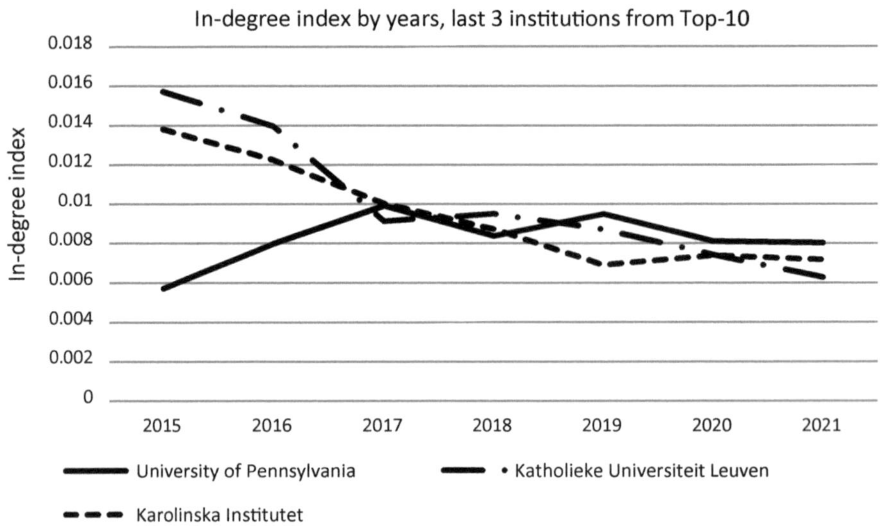

Fig. 7.10 In-degree index value by year

list—from No. 8 to No. 10 (see Table 7.16). The citation dynamics of the top three from the list (No. 1–No. 3 in Table 7.17) were shown in Fig. 7.9.

Note that while the In-degree index value for the University of Pennsylvania changed insignificantly over the entire period under review, for Katholieke Universiteit Leuven and Karolinska Institutet the In-degree index declined. Most likely, the decline in citations for these affiliations is due to a slowdown in the number of publications on the subject under consideration, while other centers increased their publication volume. Thus, the National Institutes of Health, ranking first in citation ratings, published 282 articles over seven years and increased its citation count from 12 (2015) to 1395 (2021). Karolinska Institutet published 208 articles over seven years and increased its citation count from 29 (2015) to 553 (2021).

Let us consider the Top-10 affiliations ranked by Betweenness index value. Table 7.18 presents the Betweenness index values and ranks for other classical indices for the Top-10 affiliations.

The Betweenness index rankings (Table 7.18) differ somewhat from the In-degree rankings (Table 7.16); new universities appear that are not among the most highly cited: Shanghai University, Radboud University Nijmegen in the Netherlands, and Capital Medical University in Beijing. This index highlights nodes that act as important intermediaries between other nodes, even if these research centers (nodes) are not among the most highly cited.

The dynamics of changes in the Betweenness index for the top four affiliations in the Top-10 are shown in Fig. 7.11. We included Shanghai Jiao Tong University in this group of four, as it entered the Top-10 list solely by the Betweenness index.

It should be noted that the indices of the UCL Institute of Neurology and the University of Cambridge decline from 2015 to 2019 and then begin to grow slowly.

Table 7.18 Top-10 affiliations by Betweenness

№	Institutions	Betweenness values	Rank		
			In-degree	Eigenvector	PageRank
1	University of Cambridge	0.0221	3	3	4
2	University of Oxford	0.0191	4	2	3
3	UCL Institute of Neurology	0.0179	2	4	2
4	Shanghai Jiao Tong University	0.0176	18	32	24
5	National Institutes of Health	0.0175	1	1	1
6	University College London	0.0173	5	5	5
7	Radboud University Nijmegen	0.0170	14	19	11
8	Harvard University	0.0151	7	8	7
9	Capital Medical University	0.0150	12	17	17
10	Karolinska Institutet	0.0144	10	9	9

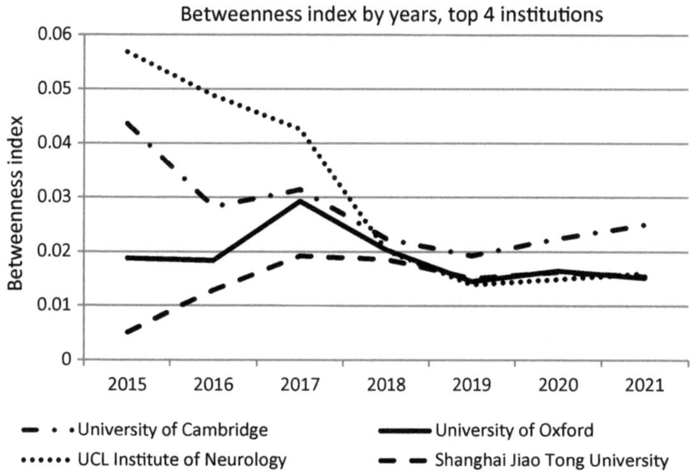

Fig. 7.11 Dynamics of changes in the Betweenness index for four top affiliations

The Betweenness index of Shanghai Jiao Tong University grows from 2015 to 2017 and then stabilizes. Shanghai Jiao Tong University increased its citation count from 7 (2015) to 583 (2021).

The Eigenvector index determines the importance of a node considering the importance of its neighbors. Table 7.19 presents the Eigenvector index values for the Top-10 affiliations.

We can see that the list of affiliations presented in Table 7.19 completely replicates the list shown in Table 7.16 with In-degree index values. Only the rank order has changed. This index assumes high values if the neighboring nodes also have high values. In this case, it is evident that prestigious universities, which cite many of each other's works, have high values, and therefore it is logical that they occupy the top of the ranked list. Figure 7.12 shows the dynamics of the Eigenvector index over the seven years considered for four affiliations. The National Institutes of Health and the University of Oxford occupy the first and second positions in the Top-10 by Eigenvector index. Karolinska Institutet and Katholieke Universiteit Leuven occupy the 9th and 10th places in the Top-10, respectively.

As can be seen from Fig. 7.12, the indices of the two leading centers—the National Institutes of Health and the University of Oxford—grow intensively from 2015 to 2020. However, in 2021, the index of the University of Oxford fell, while the index of the National Institutes of Health slowed in growth. During this period, the index of Katholieke Universiteit Leuven steadily declines, and the index of Karolinska Institutet slowly increases from 2015 to 2017, then slightly falls in 2018, and remains practically unchanged. At the same time, as already noted above, the citations of the top two affiliations in the Top-10 are nearly twice as high as those of the last two in the Top-10.

The PageRank index, like the Eigenvector index, has a similar meaning, but the importance of neighboring nodes is proportional to the number of outgoing links.

Table 7.19 Top-10 affiliations by Eigenvector

№	Institutions	Eigenvector value	Rank		
			In-degree	Betweenness	PageRank
1	National Institutes of Health	0.4312	1	5	1
2	University of Oxford	0.2865	4	2	3
3	University of Cambridge	0.2842	3	1	4
4	UCL Institute of Neurology	0.2780	2	3	2
5	University College London	0.2505	5	6	5
6	Northwestern University	0.2209	6	14	6
7	University of Pennsylvania	0.1582	8	24	8
8	Harvard University	0.1564	7	8	7
9	Karolinska Institutet	0.1290	10	10	9
10	Katholieke Universiteit Leuven	0.1276	9	26	12

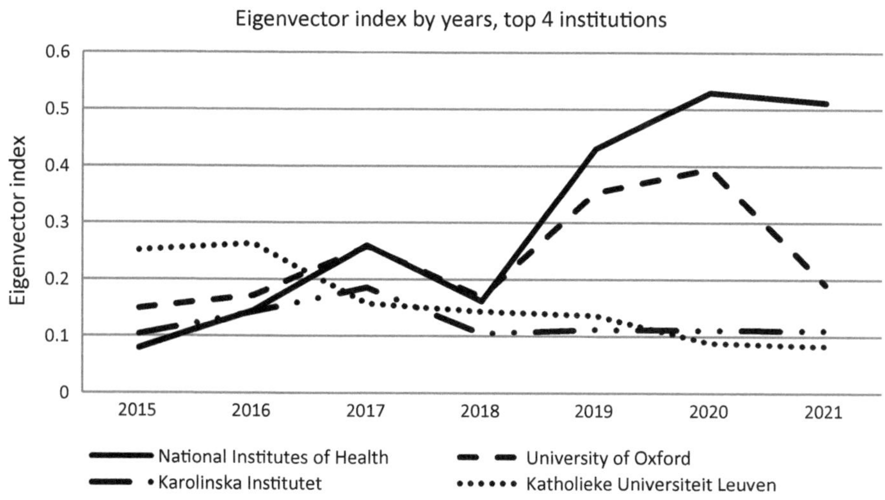

Fig. 7.12 Dynamics of changes in the Eigenvector index for four affiliations

The Top-10 affiliation list by PageRank index (Table 7.20) differs from the list in Table 7.19 by replacing Katholieke Universiteit Leuven with King's College London (No. 10). King's College London is part of the University of London, as is the UCL Institute of Neurology (No. 2 on the list).

As an example, Fig. 7.13 shows the dynamics of the PageRank index for the top three affiliations from the Top-10 and King's College London.

The dynamics of the PageRank index in 2015–2021 for King's College London shows that the index value remains almost constant without sharp changes.

Next, we present the results of a network analysis of new centrality indices—the Bundle Index (BI), the Pivotal Index (PI), and the Total Influence Index (TI).

7.6.2 *New Centrality Indices*

The following parameters were selected for the network analysis using new centrality indices: $k = 3$—the size of the group of nodes (critical set) that can simultaneously exert influence, and several quota values: $q = 0.1\%, 0.5\%, 1\%, 3\%, 5\%, 10\%$.

Since authors of publications can cite both their own works and those of colleagues working in the same organization, loops appear in the affiliation network. In this network, a node may belong to its own critical set.

Table 7.21 presents the BI, PI, and TI index values with the smallest quota $q = 0.1\%$ for the Top-10 affiliations. The list is ranked by TI value. For comparison, the table also includes In-degree index values for the affiliations.

As seen from Table 7.21, the Top-10 affiliation list for quota $q = 0.1\%$ differs little from the Top-10 list by In-degree index (Table 7.16). However, a different ranking

Table 7.20 Top-10 affiliations by PageRank

№	Institutions	PageRank values	Rank		
			In-degree	Betweenness	Eigenvector
1	National Institutes of Health	0.0125	1	5	1
2	UCL Institute of Neurology	0.0117	2	3	4
3	University of Oxford	0.0114	4	2	2
4	University of Cambridge	0.0107	3	1	3
5	University College London	0.0100	5	6	5
6	Northwestern University	0.0095	6	14	6
7	Harvard University	0.0091	7	8	8
8	University of Pennsylvania	0.0076	8	24	7
9	Karolinska Institutet	0.0072	10	10	9
10	King's College London	0.0067	11	17	11

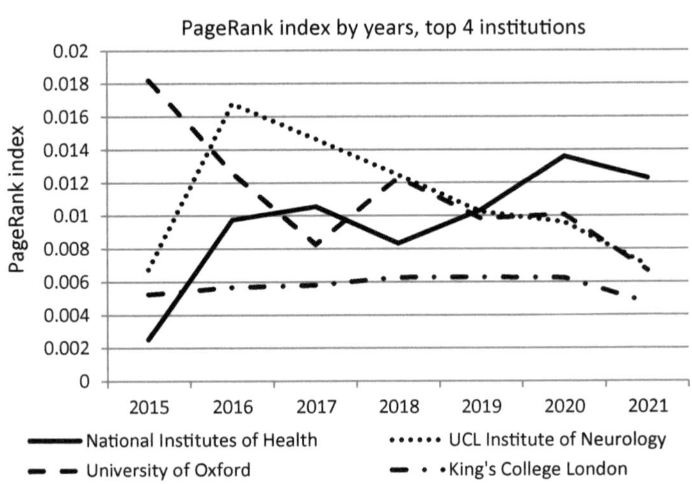

Fig. 7.13 Dynamics of changes in the PageRank index for four affiliations

order can be observed. Furthermore, instead of Katholieke Universiteit Leuven, Capital Medical University appears in the list, which also ranked in the Top-10 by Betweenness (No. 9 in Table 7.18). The similarity between the In-degree and new indices lists with a small quota ($q = 0.1\%$) is explained by the fact that at small

Table 7.21 BI, PI, and TI index values for the Top-10 affiliations, quota $q = 0.1\%$

№	Institutions	In-degree	$q = 0.1\%$		
			BI	PI	TI
1	UCL Institute of Neurology	0.0134	0.0436	0.2107	0.0892
2	National Institutes of Health	0.0156	0.0365	0.1600	0.0707
3	University of Cambridge	0.0131	0.0321	0.1390	0.0614
4	University College London	0.0120	0.0367	0.1054	0.0514
5	Harvard University	0.0100	0.0331	0.1088	0.0506
6	University of Oxford	0.0122	0.0312	0.1029	0.0487
7	Northwestern University	0.0114	0.0306	0.0972	0.0464
8	University of Pennsylvania	0.0085	0.0193	0.0685	0.0321
9	Capital Medical University	0.0078	0.0225	0.0004	0.0102
10	Karolinska Institutet	0.0079	0.0198	0.0003	0.0093

quota values the number of critical sets for nodes remains sufficiently large; it is not difficult to exceed the quota for groups of citing nodes, hence the results are similar to the In-degree index.

To illustrate the yearly dynamics of the indices, we use the example of TI index values ($q = 0.1\%$) for Capital Medical University, which entered the Top-10 by the Betweenness index, and UCL Institute of Neurology—a leader by all indices. We chose the TI index since it is a linear combination of the Bundle Index (BI), the Pivotal Index (PI), and the In-degree index. Figure 7.14 shows the graph of TI index values ($q = 0.1\%$) for each year under review. The TI index of the UCL Institute of Neurology, having peaked in 2016, declines in subsequent years. The same trend was observed for this affiliation with classical indices (see Figs. 7.11–7.13). For Capital Medical University, the index remains nearly unchanged year by year.

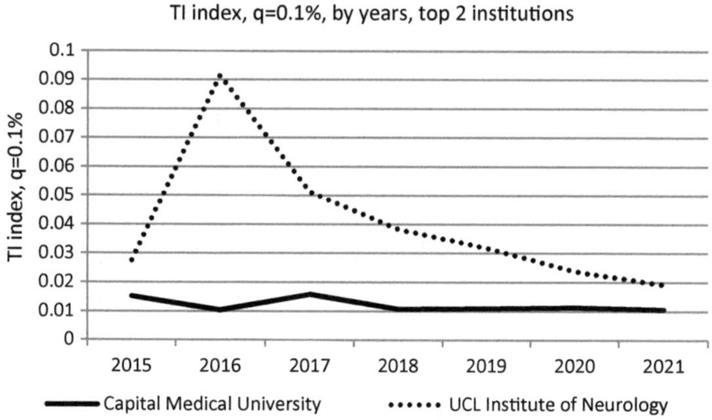

Fig. 7.14 Dynamics of changes in the TI index for two affiliations

Including loops (self-citations) in the affiliation network and the different numbers of citations between affiliations showed that quotas up to 5% almost do not change the Top-10 affiliation list. Table 7.22 presents BI, PI, and TI index values with a quota of $q = 5\%$ for the Top-10 affiliations. For comparison, the table also provides affiliation ranks by these indices and by the In-degree index. The list is ranked by TI value.

From Table 7.22, it can be seen that the Top-10 affiliation lists by TI, BI, and PI indices with quota $q = 5\%$ differ only in position in the ranked list and the rank of the University of Pennsylvania, which dropped out of the top ten. Position No. 10 in the BI index ranked list is held by Karolinska Institutet. According to the PI index, Université de Montréal ranks eighth. Comparison with the Top-10 affiliations by the In-degree index showed that Katholieke Universiteit Leuven and Karolinska Institutet dropped out of the top ten, while King's College London and Capital Medical University, previously ranked 11th and 12th in the In-degree index list, entered the Top-10. Comparing the Top-10 affiliation lists by TI, BI, and PI indices with quotas $q = 0.1\%$ (Table 7.21) and $q = 5\%$ (Table 7.22), one can observe minor differences. Karolinska Institutet (No. 10 in Table 7.21) dropped out of the top ten, and King's College London under No. 4 (for TI) entered the Top-10 (Table 7.22). It is worth noting that King's College London in the Top-10 lists by BI and PI with quota $q = 5\%$ ranks first and second, respectively.

The only quota of $q = 10\%$ for the indices TI, BI, PI resulted in more differences with In-degree in the Top-10 lists. Table 7.23 shows the Top-10 affiliation ranks by the indices TI, BI, PI with a quota of $q = 10\%$ and In-degree. Affiliations are ranked by TI.

In the Top-10 list by the PI index with a quota of $q = 10\%$, the University of Padua is ranked 5-th. In the Top-10 list by the BI index, it took the 14th position. Comparing the Top-10 lists by the In-degree index (Table 7.16) and the indices TI, BI, PI, we see that five out of ten affiliations (No. 4, 6, 7, 8, 10; see Table 7.23) did not

Table 7.22 BI, PI, and TI index values for the Top-10 affiliations, quota $q = 5\%$

№	Institutions	In-degree	$q = 5\%$		
			BI	PI	TI
1	UCL Institute of Neurology	0.0134	0.0436	0.2107	0.0892
2	National Institutes of Health	0.0156	0.0365	0.1600	0.0707
3	University of Cambridge	0.0131	0.0321	0.1390	0.0614
4	University College London	0.0120	0.0367	0.1054	0.0514
5	Harvard University	0.0100	0.0331	0.1088	0.0506
6	University of Oxford	0.0122	0.0312	0.1029	0.0487
7	Northwestern University	0.0114	0.0306	0.0972	0.0464
8	University of Pennsylvania	0.0085	0.0193	0.0685	0.0321
9	Capital Medical University	0.0078	0.0225	0.0004	0.0102
10	Karolinska Institutet	0.0079	0.0198	0.0003	0.0093

Table 7.23 Top-10 ranks of affiliations by TI, BI, PI with a quota of q = 10% and In-degree

TI rank	Institutions	Rank		
		In-degree	BI	PI
1	National Institutes of Health	1	1	1
2	University of Cambridge	3	2	2
3	University of Oxford	4	3	3
4	Capital Medical University	12	4	4
5	University of Pennsylvania	8	6	7
6	Radboud University Nijmegen	14	5	6
7	Shanghai Jiao Tong University	18	7	8
8	University of Florida	21	8	9
9	Katholieke Universiteit Leuven	9	9	10
10	Newcastle University	16	10	11

appear in the Top-10 by In-degree. The five affiliations that were previously in the In-degree Top-10—UCL Institute of Neurology (No. 2, Table 7.16), University College London, Northwestern University, Harvard University (No. 5, 6, and 7, Table 7.16), and Karolinska Institutet (No. 10, Table 7.16)—have now fallen out of the Top-10.

The dynamics of changes in the TI index with a quota of $q = 10\%$ from 2015 to 2021 are shown for three affiliations: Radboud University Nijmegen, University of Florida, and Newcastle University (Fig. 7.15).

It should be noted that the TI index with a quota of $q = 10\%$ for these three affiliations showed growth from 2015 to 2020. In 2021, the TI index for Radboud University Nijmegen and University of Florida continued to rise. However, the TI index for Newcastle University decreased nearly threefold, from 0.014 to 0.005.

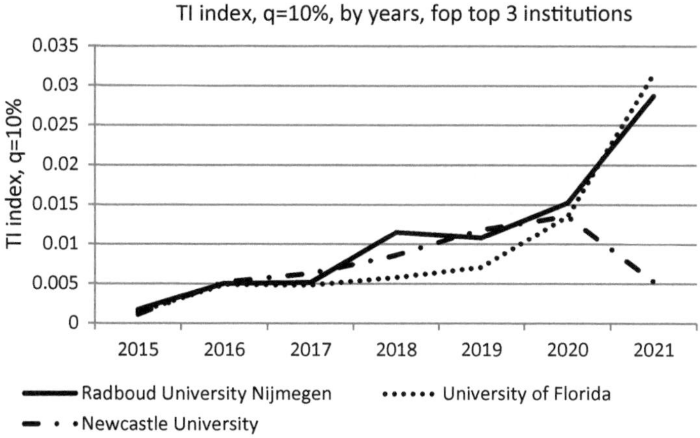

Fig. 7.15 Dynamics of the TI index for three affiliations

Table 7.24 Top-10 affiliations by TI with a quota of $q = 10\%$

№	Institutions	In-degree rank	Number of citations	Share of self-citations
1	National Institutes of Health	1	3757	0.112
2	University of Cambridge	3	3165	0.125
3	University of Oxford	4	2933	0.122
4	Capital Medical University	12	1878	0.111
5	University of Pennsylvania	8	2046	0.010
6	Radboud University Nijmegen	14	1799	0.125
7	Shanghai Jiao Tong University	18	1677	0.103
8	University of Florida	21	1573	0.122
9	Katholieke Universiteit Leuven	9	1927	0.107
10	Newcastle University	16	1702	0.147

The analysis of new indices with a higher quota revealed differences from the article network. In the article network, as the quota increased, the number of critical sets decreased, and articles with the highest citation counts had low values for the new indices. Thus, articles that were actively cited in narrowly specialized groups, but did not receive many citations in the broader community, became significant. In the affiliation network, due to the presence of loops and the assumption that a vertex can belong to the critical set for itself, even for larger quotas, the most cited affiliations remain in the Top-10 (Table 7.24). This occurs because these affiliations have a sufficiently large percentage of self-citations, which allows sets with these vertices to become critical and exceed the threshold.

7.6.3 *An Analysis of Citation Network Stability for Affiliations*

The constructed networks can change their structure over time. New vertices appear, old ones disappear, and relationships between them change. It is important to understand how much the network has changed over time in order to identify patterns and trends. In our research, we used a method of network stability analysis based on comparing networks by topology and vertex centrality.

The main task of this method is to compare the same network over different years. Networks are compared using two different methods: comparing changes

Table 7.25 Stability of the citation network for affiliations

	2015–16	2016–17	2017–18	2018–19	2019–20	2020–21
Distance between interval order matrices	0.0555	0.0554	0.0422	0.0581	0.0541	0.0523
Distance between influence matrices	0.00016	0.00023	0.00029	0.00034	0.00039	0.00044
Stability	0.03921	0.03916	0.02983	0.04108	0.03826	0.03697

in vertex centrality and comparing changes in network structure. The change in vertex centrality is assessed using a matrix that shows relative differences in vertex centrality. For each year under consideration, interval order matrices and influence matrices for consecutive years were calculated. Then, the distance between two adjacent matrices was calculated. When calculating, a change in the rank of an affiliation by less than 10 units was considered insignificant, i.e., when $\varepsilon \leq 10$, the ranks were considered equal. The calculated data are shown in Table 7.25.

The distance between the interval order matrices is close to zero and changes slightly over time, which indicates minor changes in the ranking of affiliations over the years. This explains the stable positions of the most important affiliations. The distance between the influence matrices is even smaller. This means that organizations cite each other in similar proportions every time. Therefore, the final values of the stability measure are also close to 0 and change little over time. Thus, affiliations' citations remain constant, and there were no significant trends that could change the citation structure over time. Therefore, when selecting potential organizations for investment, one can rely on their recognition and importance in previous years.

7.7 Clustering of Affiliation Patterns by Changes in Citation Rates Over the years

For determining changes in the citation count of affiliations from 2015 to 2021, affiliation clustering and identification of differing patterns were conducted. Pattern analysis was carried out based on changes in the In-degree index over the years. Methods were applied using a fixed tube for absolute values and tangents of slope angles, with the tube width (ε) varying from 0.05 to 0.3. Data were normalized using a min–max approach by year (see Chap. 2). As an example, we consider the clustering of the In-degree index using tangents with the parameter $\varepsilon = 0.1$.

For other parameters and clustering methods, the results were similar.

The analysis of pattern grouping for all indices revealed a general trend. The obtained patterns can be divided into 3 groups:

1. More than 90% of affiliations with a small number of citations are grouped into one pattern and show insignificant changes in indices.

2. Patterns of affiliations with high index values, which are often ranked at the top of the centrality-index rankings, or those with the most pronounced changes, are unique. These patterns contain only one affiliation.
3. Clustering of 3–5 affiliations that have similar patterns.

The given parameters for clustering the In-degree index resulted in 29 patterns, the largest of which contained 2833 organizations, while 18 patterns contained one organization each. Below are some examples.

First group.

Pattern 1 included 2833 affiliations, representing 93.5% of the total number of affiliations. Changes were insignificant, primarily a slight upward followed by a downward trend. The majority of affiliations had small index values and were located at the bottom of the ranked list for the entire period.

Pattern 2—106 affiliations with a small number of citations and a very weak upward trend.

Pattern 3—32 affiliations with small values and a declining citation trend from 2015 to 2017. In subsequent years (2018–2021), weak changes in citations were observed.

Pattern 4 included 16 affiliations, mostly not in the Top-100.

Second group of patterns primarily included affiliations ranked at the top of the centrality index lists. The graphical representation of the affiliations forming a separate pattern helps understand how a distinct pattern might emerge. For example, it helps identify affiliations that were not significant throughout the entire period but had maximum values in a particular year. As an example, see Pattern No. 14—Northwestern University in Fig. 7.16.

Northwestern University was ranked 6-7th in terms of all index values, but when considering the pattern, it is evident that the number of references gradually increased and reached its maximum in 2019, after which there was a sharp decline.

Another example—Pattern 29—National Institutes of Health (NIH) (Fig. 7.17).

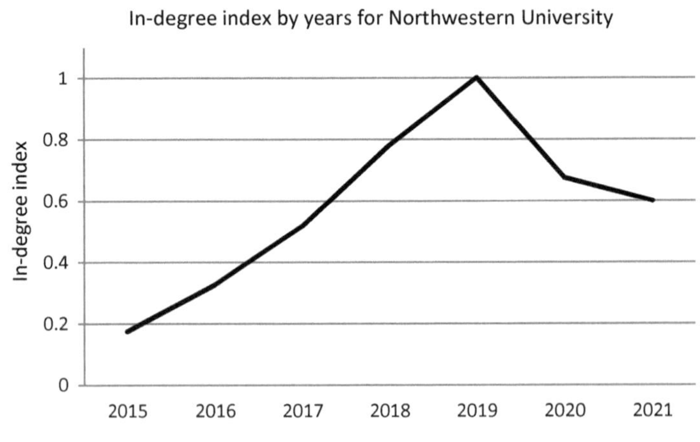

Fig. 7.16 Pattern No. 14—Northwestern University

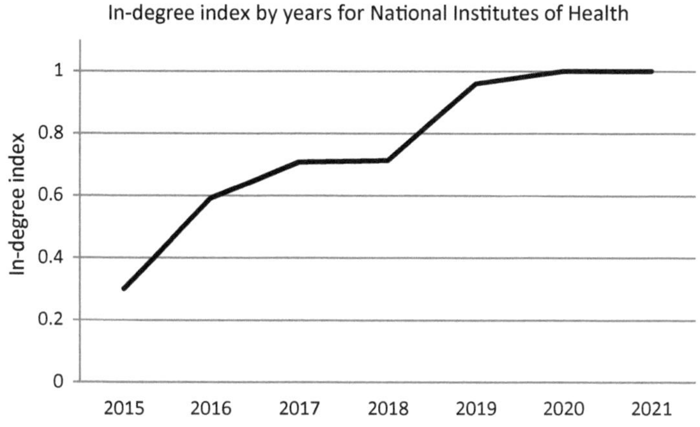

Fig. 7.17 Pattern No. 29—National Institutes of Health (NIH)

NIH consists of 27 institutes and research centers and is the main U.S. government center responsible for research in healthcare and biomedicine. It had the highest number of citations in 2015–2021.

We see a steadily increasing trend, but the institute took first place only in 2020 and 2021.

Pattern 6 consisted of 4 institutions: two universities from the USA (Harvard University—ranked 7th by citations, and University of Alabama at Birmingham—ranked 17th), University of Padua in Italy (ranked 43rd), and Radboud University Nijmegen in the Netherlands (ranked 14th). All universities had an upward citation trend until 2019, after which the trend declined (Fig. 7.18).

The third group of patterns is the most interesting because it reveals intriguing regularities. For instance, in Fig. 7.19, Pattern No. 9 includes institutions from one

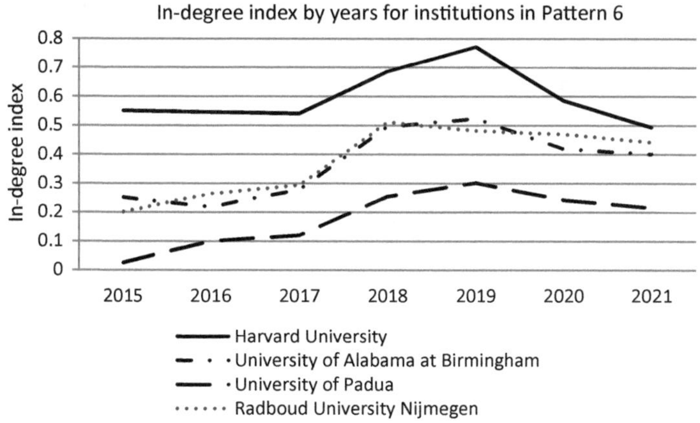

Fig. 7.18 Pattern No. 6

country with very similar citation dynamics (University of Cambridge and University College London, part of the University of London).

Certain patterns may indicate regularities in dynamic changes. Figure 7.20 and 7.21 show patterns with increasing and decreasing trends. While there are small fluctuations from year to year, such patterns generally allow for the assessment of the potential of an affiliation in the research area.

The University of Barcelona (Fig. 7.20) was ranked 2nd by citations in 2015 (37 references) for articles on the studied topic. Despite the number of citations increasing from 2015 to 2021, its citation ranking dropped. In 2021, the university ranked 27th (393 references). The overall ranking of the University of Barcelona over seven years was 26th (1304 references). This can be explained by the fact that 122 articles on the studied topic were published during this period. In comparison,

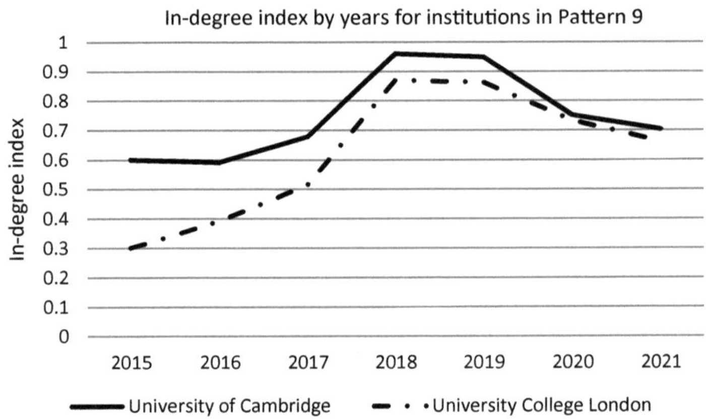

Fig. 7.19 Pattern No. 9

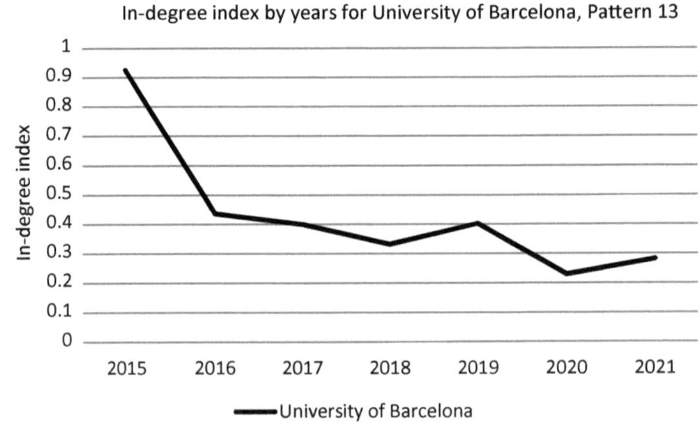

Fig. 7.20 Pattern No. 13

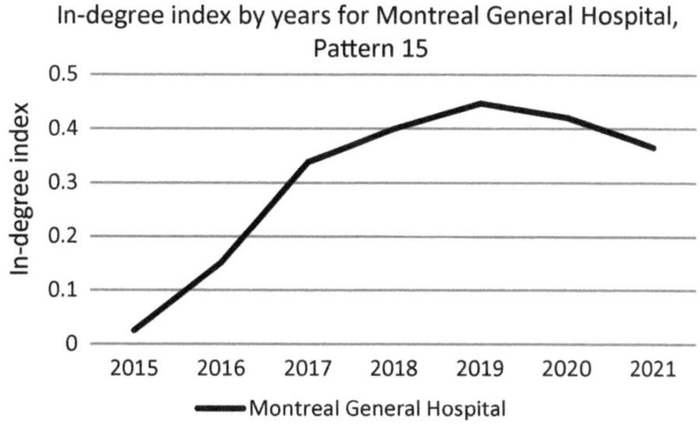

Fig. 7.21 Pattern No. 15

UCL Institute of Neurology, ranked 2nd in citation rankings, published 206 articles (3233 references) during the same period. Another factor is the decreasing interest in the topic. For example, Katholieke Universiteit Leuven, ranked 9th by citations, published 124 articles, which had 1927 citations.

Montreal General Hospital (Pattern 15, Fig. 7.21) ranks 22nd in citation ranking. From 2015 to 2021, the number of citations increased from 1 to 509. The total number of citations for 8 publications on Parkinson's Disease (PD) is 1559.

Thus, pattern analysis allows identifying interesting regularities in the trends of parameter changes for the objects under consideration. It enables the analysis of objects that may not always be highlighted through centrality analysis and presents them in a convenient form.

7.8 An Analysis of Independent Citation Networks of Affiliations

To exclude artificial enhancement of citation counts for organizations, a model of independent citations of affiliations was developed. This model is a citation graph of affiliations, which considers all pairwise citations of affiliations, excluding edges that cite their own work or co-authors' works. If paper ID1 is written by 5 authors, and paper ID2 is written by 6 authors, all possible citations between authors of ID1 and ID2 can occur, up to a maximum of 30 in this example. Citations of authors to themselves and citations between authors who have been co-authors at any time within the considered period are excluded. Then, a citation network of affiliations for these authors is constructed from the remaining citations, with only unique edges considered for each paper, and the weight set to 1. These resulting edges are called independent citations. The described steps are repeated for all citations. This model helps exclude cases of artificial citations and assess the real significance of affiliations.

Table 7.26 The characteristics of the affiliations network

Name of the indicator	Value
Number of nodes	4682
Number of edges	670,890
Number of connected components	2
Largest connected component	4681
Network density	0.03
Minimum number of citations	1
Maximum number of citations	3711
Average number of citations	286.6

This network includes more affiliations than the previous one because it considers all affiliations, not just the first author. The characteristics of the affiliations network are presented in Table 7.26.

7.8.1 *Centrality Analysis*

In the constructed network, many new organizations appear among the most cited ones. This indicates that one particular affiliation frequently occupies the top position, while some co-author affiliations rarely appear at the top. For example, the top-20 includes Johns Hopkins University School of Medicine, German Center for Neurodegenerative Diseases, University of Tuebingen, University of Washington, University of California, Genentech. These organizations are from Germany and the USA, some of them are universities not specializing in medicine, while others focus on specialized research. Compared to the citation network of the first affiliation, their positions improved by 20–80 points, highlighting their significant role as co-author affiliations.

A comparison was also made between the complete citation network of affiliations (which did not exclude co-author citations) and the independent citation network. The initial analysis showed a difference in ranking by citation count. To numerically assess the differences, a stability measure was used for the resulting networks. When comparing the two networks by years with $\varepsilon = 5$, the values of distances were less than 0.001. The differences are small, arising from the ranking differences, but the overall structure of interactions and influences between vertices did not change. It can be assumed that artificial citation enhancement of co-authors is not practiced in this community. Mostly, co-author citations occur within the same organization. Table 7.27 presents the five organizations with the highest number of co-author citations from all organizations.

Table 7.28 presents the Top-5 organizations with works citing co-authors, from the hundred most cited. Despite the networks with and without self-citations being

Table 7.27 Top-5 affiliations by co-author citations

№	Institution	Number of citations of co-authors
1	National Institutes of Health	26,322
2	University of Cambridge	17,870
3	Central South University	14,520
4	Juntendo University	11,897
5	Johns Hopkins University School of Medicine	11,082

Table 7.28 Top-5 affiliations citing co-authors

№	Institutions	Number of citations of co-authors	Share of citations of co-authors	Change in rank
1	Montreal Neurological Institute and Hospital	41,575	0.366	−16
2	McGill University	44,095	0.350	−13
3	Oslo University Hospital	14,197	0.337	−21
4	National Institutes of Health	106,316	0.275	0
5	Iowa State University	12,170	0.270	+ 47

close in stability, changes in the network are still observed. This is especially noticeable among medium-citation organizations, whose rank changes by more than 10 points. Since the metric only considers the presence of rank changes, the networks are considered similar due to the large number of nodes that retain their relative position. Looking at the absolute rank change, it averages 32, which is significant, so considering the independent network is meaningful.

7.9 An Analysis of the Network of Organizational Collaboration

The network of organizational collaboration is an undirected graph, where the vertices represent the affiliations of authors, and the edges exist if the authors of these affiliations have collaborated, i.e., if they are co-authors. The weight of an edge corresponds to the number of joint papers written in a given year. Such a network may contain loops if the paper is authored by individuals from the same institution. Also, if an author has multiple affiliations listed, all are considered in the network. Isolated affiliations, which do not interact with other affiliations, were not removed from the network. Since each paper contains from 1 to 436 authors and from 1 to 138 affiliations, the resulting network is much larger than previous ones.

Network Attributes:

- AfId1—ID of one author

- AfId2—ID of another author
- Y—Year of joint publication of the authors
- weight—Number of joint publications between affiliations AfId1 and AfId2 in the year Y, i.e., the weight of the edge

Author ID is a numeric identifier assigned by the Microsoft Academic search engine.

The organizational collaboration network is built based on all author affiliations, meaning that it involves more affiliations than previous networks, although its density is lower. The network has the following parameters:

- 4697 vertices, of which 63 are isolated
- 136,405 edges (or 85,900 unique edges, not accounting for the year)
- 270 connected components, with 4396 being the size of the largest, and the others containing 1–3 affiliations
- Density: 0.008
- Minimum number of connections: 0
- Maximum number of connections: 814
- Average number of connections: 36.6

7.9.1 Centrality Analysis

In the interaction network, the degree of a vertex corresponds to the number of different organizations with which joint work has been done. Table 7.29 shows the Top-10 organizations by the number of connections. In an undirected network, it is already apparent that the rank by Betweenness is similar to the rank by vertex degree, meaning that the greatest number of paths passes through a vertex with many connections. Interestingly, organizations with more connections do not always have high citation counts. The organizations from the Top-10 by citations that are not included in this list interact with fewer universities, ranging from 232 to 576. This may indicate that these organizations are more selective in choosing partners.

7.9.2 Dynamic Analysis

For the dynamic analysis of the structure of author affiliation interactions, network stability (Table 7.30) and parameters presented in Table 7.31 were calculated.

This network is less stable compared to the citation network of affiliations, and it is evident that the values increase over time. Significant changes are observed in the interval orders, i.e., many organizations change their positions in the ranked list due to changes in the number of connections.

Overall, there is an increasing trend in the number of vertices, i.e., the number of organizations engaged in research in the field of PD is growing and expanding.

Table 7.29 Top-10 affiliations by the number of connections

№	Institutions	Number of connections	Betweenness rank	In-degree rank
1	Harvard University	814	1	7
2	National Institutes of Health	784	2	1
3	University of Cambridge	726	3	3
4	University College London	725	4	5
5	King's College London	677	5	11
6	University of Pennsylvania	629	6	8
7	Karolinska Institutet	627	9	10
8	Mayo Clinic	620	8	13
9	University of Toronto	600	7	33
10	German Center for Neurodegenerative Diseases	599	13	37

Table 7.30 Stability of the organizational collaboration network

	2015–2016	2016–2017	2017–2018	2018–2019	2019–2020	2020–2021
Distance between interval order matrices	0.112	0.131	0.129	0.127	0.143	0.151
Distance between influence matrices	0.00014	0.00015	0.00016	0.00016	0.00018	0.00019
Stability	0.0791	0.0928	0.0912	0.0895	0.101	0.107

The number of isolated affiliations not participating in joint research does not exceed 0.7%, indicating a high degree of collaboration in this field. This is also confirmed by the sizes of the connected components: the largest contains nearly 90% of the vertices. In 2019, a noticeable increase in components, a decrease in participants in the largest group, a decrease in graph density, and an increase in new organizations and interactions between them were observed. This may indicate that, after several years, new papers were published by certain co-author groups who had not been actively working with the rest of the community. Despite this, by 2021 the number of components decreased, and more organizations joined the main group. The average degree of a vertex fluctuated over 7 years, peaking in 2017 and increasing by 6 by 2021 compared to 2015. It can be concluded that organizations are expanding their circle of interactions. The share of new edges and new vertices is gradually decreasing, i.e.,

Table 7.31 Characteristics of the organizational collaboration network by year

Metric	2015	2016	2017	2018	2019	2020	2021
Number of nodes	1799	1980	2129	2110	2316	2541	2654
Number of edges	11,440	16,901	23,193	15,996	17,864	26,031	24,980
Number of isolated nodes	12	6	4	7	12	14	12
Proportion of isolated nodes	0.007	0.003	0.002	0.003	0.005	0.006	0.005
Number of connection components	188	188	188	182	214	210	197
Size of largest connection component	1564	1731	1892	1884	2029	2280	2402
Proportion of nodes in the largest component	0.870	0.874	0.889	0.893	0.876	0.897	0.905
Size of second largest connection component	4	7	4	4	5	5	6
Average node degree	12.72	17.07	21.79	15.16	15.43	20.49	18.82
Network density	0.007	0.009	0.010	0.007	0.007	0.008	0.007
Proportion of new nodes	–	0.365	0.272	0.176	0.181	0.172	0.140
Proportion of new edges	–	0.795	0.748	0.574	0.503	0.524	0.475
Proportion of remaining nodes	–	0.635	0.644	0.686	0.652	0.630	0.687
Proportion of remaining edges	–	0.205	0.188	0.287	0.313	0.271	0.344
Proportion of newly appeared nodes	–	–	0.084	0.138	0.167	0.168	0.173
Proportion of newly appeared edges	–	–	0.065	0.140	0.184	0.205	0.181

the circle of research centers involved in PD is approaching a constant. Meanwhile, the increasing share of remaining vertices and edges indicates a strengthening of interest from universities in the field of PD research and collaboration with other organizations. The increase in returning connections also shows strong interest in the field.

7.10 Authors and Affiliations

By analyzing the affiliation and author networks, we noted that the most cited authors are not always employees of organizations with a large number of citations. Conversely, a high number of citations for an affiliation does not correlate with the citation count of its employees. As an example, Table 7.32 presents the Top-20 affiliations ranked by the In-degree index, the number of citations for these affiliations from 2015 to 2021, and the authors' ranks according to the In-degree index for authors with these affiliations.

Table 7.32 Top-20 affiliations and authors ranked by In-degree

Inst. In-degree rank	Institution	Total citations	Author ID	Auth. In-degree rank
1	National Institutes of Health	3757	1761635969	4
			2236297047	21
			2111276134	110
			2160055852	194
			2600996391	6
2	UCL Institute of Neurology	3233	2099613546	18
			2191234707	31
			2551724242	102
			1804868078	100
			2012590864	105
			2308062378	111
3	University of Cambridge	3165	2138533061	28
			1467683049	29
			2116447771	57
			2120275278	83
			2162721967	135
4	University of Oxford	2933	2748676791	45
			2504096529	116
			2128437711	124
			2530150922	169
5	University College London	2882	2064657699	2
			2618592063	52
			2191234707	31
			2140665068	96
			2304566372	130
			2105050807	131
6	Northwestern University	2759	378198483	13
			1256124130	27
			2731932222	46
			328685278	58
			1878031441	178
7	Harvard University	2413	1991986455	19
			2166127851	89
			2066071986	231
8	University of Pennsylvania	2046	2309451887	75
			2233079596	154

(continued)

Table 7.32 (continued)

Inst. In-degree rank	Institution	Total citations	Author ID	Auth. In-degree rank
9	Katholieke Universiteit Leuven	1927	2265103992	12
10	Karolinska Institutet	1893	2599128148	37
			3183379636	40
11	King's College London	1885	167518267	23
			2049643778	60
			2557672825	77
			2094971281	196
			2251562891	188
12	Capital Medical University	1878	2539383013	189
13	Mayo Clinic	1863	2145570510	164
			40383365	163
			1929534198	165
14	Radboud University Nijmegen	1799	1912620772	86
			2581295567	122
15	Newcastle University	1702	2111253536	64
16	University of Alabama at Birmingham	1694	1898612074	103
17	Shanghai Jiao Tong University	1677	2508761563	280
18	University of Sydney	1592	2130269896	159
			2524178533	138
			2055023048	152
			2124696176	180
19	University of Florida	1573	2408963588	57
			2984353310	119
			1913936031	147
20	Montreal General Hospital	1559	2047227543	1

Table 7.33 Top-20 authors and affiliations ranked by In-degree

Author ID	In-degree rank of the author	Affiliation	In-degree rank of the affiliation
2047227543	1	Montreal General Hospital	22
2064657699	2	University College London	5
2303827053	3	University of Rochester	78
1761635969	4	National Institutes of Health	1
78852038	5	University of Cincinnati	72
2600996391	6	National Institutes of Health (Biochemistry Section. Surgical Neurology Branch. National Institute of Neurological Disorders and Stroke (NINDS))	1
2570671491	7	Max Planck Society	40
2100557140	8	Veterans Health Administration	85
2153866458	9	University of Padua	43
2175448217	10	German Center for Neurodegenerative Diseases	37
2112352720	11	California Institute of Technology	107
2265103992	12	Katholieke Universiteit Leuven	9
378198483	13	Northwestern University	6
2084545881	14	University of Helsinki (Central Hospital Department of Neurological Sciences)	63
2127596906	15	Toronto Western Hospital	31
2195539447	16	Haukeland University Hospital	81
2026404387	17	Tel Aviv Sourasky Medical Center	124
2099613546	18	UCL Institute of Neurology	2
1991986455	19	Harvard University	7
2154075639	20	Imperial College London	41

Notice that this list includes only 4 authors from the Top-10. The author with AuId 2047227543, which has rank 1 (Montreal General Hospital), two authors from the National Institutes of Health—AuId 1761635969 (rank 4) and 2600996391 (rank 6), and the author AuId 2064657699 (rank 2) from University College London. Another four authors are listed in the Top-20.

Table 7.33 presents the Top-20 authors ranked by In-degree, with their affiliations and affiliation ranks indicated.

In this list of 20 authors, only six affiliations belong to the Top-10.

Thus, the conclusion that the correlation between the leading affiliations by citation number and authors with a high number of citations is low has been confirmed.

References

Aria, M., & Cuccurullo, C. (2017). Bibliometrix: An R-tool for comprehensive science mapping analysis. *Journal of Informetrics, 11*(4), 959–975. https://doi.org/10.1016/j.joi.2017.08.007

Chen, Q., Sun, T., & Wang, T. (2023). Network centrality, support organizations, exploratory innovation: Empirical analysis of China's integrated circuit industry. *J Heliyon, 9*(7), Article e17709. https://doi.org/10.1016/j.heliyon.2023.e17709

Ding, J., Liu, C., & Kandonga, G. A. (2020). Exploring the limitations of the h-index and h-type indexes in measuring the research performance of authors. *Scientometrics, 122*, 1303–1322. https://doi.org/10.1007/s11192-020-03364-1

Duggan, E. W., Atwood, G. S., Sanford, J. A., Tsai, M. H., Egbaria, J. K., Carmichael-Tanaka, N., & Outland, N. B. (2024). Using bibliometric data to define and understand publishing network equity in anesthesiology. *Anesthesia & Analgesia, 139*(5), 944–954. https://doi.org/10.1213/ANE.0000000000006877

Ho, Y. S., & Shekofteh, M. (2021). Performance of highly cited multiple sclerosis publications in the Science Citation Index expanded: A scientometric analysis. *Multiple Sclerosis and Related Disorders, 54*, Article 103112. https://doi.org/10.1016/j.msard.2021.103112

Shishehgarkhaneh, M. B., Keivani, A., Moehler, R. C., Jelodari, N., & Laleh, S. R. (2022). Internet of Things (IoT), Building Information Modeling (BIM), and Digital Twin (DT) in construction industry: A review, bibliometric, and network analysis. *Buildings, 12*(10), 1503. https://doi.org/10.3390/buildings12101503

Sinha, A., Shen, Z., Song, Y., Ma, H., Eide, D., Hsu, B. -J., & Wang, K. (2015). An overview of Microsoft Academic Service (MAS) and applications. *Proceedings of the 24th International World Wide Web Conferences*, 243–246. https://doi.org/10.1145/2740908.2742839

Wang, K., Shen, Z., Huang, C., Wu, C.-H., Eide, D., Dong, Y., Qian, J., Kanakia, A., Chen, A., & Rogahn, R. (2019). A review of Microsoft Academic Services for science of science studies. *Frontiers in Big Data, 2*, 45. https://doi.org/10.3389/fdata.2019.00045

Yeung, A. W. K., Litvinova, O., Bragazzi, N. L., et al. (2024). Digital health and mobile health: A bibliometric analysis of the 100 most cited papers and their contributing authors. *Explor Digit Health Technol, 2*, 86–100. https://doi.org/10.37349/edht.2024.00013

Zhou, H., Zhu, X., Dai, J., & Wu, W. (2023). Innovation evolution of industry-university-research cooperation under low-carbon development background: In case of 2 carbon neutrality technologies. *Sustainable Energy Technologies and Assessments, 55*, Article 102976. https://doi.org/10.1016/j.seta.2022.102976

Chapter 8
Conclusion

Historically, the impact of any article, author or journal had been evaluated by a scientific community, sometimes too late. Today, due to enormous increase of the number of publications there is a need in high quality quantitative models analyzing the impact of a journal or an institution or an author to the corresponding field of research.

Responding to this demand many colleagues suggested some models based on very different tools—from aggregation models to semantic analysis, from statistics to optimization techniques.

We presented an attempt to construct such models based on a new technique proposed for social network analysis. This very technique include the new centrality measures, models of pattern analysis and stability.

These models were tested on the data related to the publications on Parkinson's Disease on the period 2015–2021.

The obtained results show that the models presented in the book can be successfully used in different fields of bibliometric analysis.

© The Editor(s) (if applicable) and The Author(s), under exclusive license to Springer Nature Switzerland AG 2026

F. Aleskerov et al., *Bibliometric Analysis by Network Models*, Contemporary Systems Thinking, https://doi.org/10.1007/978-3-032-09171-0_8

Index

© The Editor(s) (if applicable) and The Author(s), under exclusive license to Springer Nature Switzerland AG 2026
F. Aleskerov et al., *Bibliometric Analysis by Network Models*, Contemporary Systems Thinking, https://doi.org/10.1007/978-3-032-09171-0

MIX
Papier aus verantwortungsvollen Quellen
Paper from responsible sources
FSC® C105338

If you have any concerns about our products,
you can contact us on
ProductSafety@springernature.com

In case Publisher is established outside the EU,
the EU authorized representative is:
Springer Nature Customer Service Center GmbH
Europaplatz 3, 69115 Heidelberg, Germany

Printed by Libri Plureos GmbH
in Hamburg, Germany